D0801832

Bozo Sapiens

BY THE SAME AUTHORS

Chances Are . . . : Adventures in Probability

BOZO SAPIENS

WHY TO ERR IS HUMAN

Michael Kaplan and Ellen Kaplan

BLOOMSBURY PRESS
New York   Berlin   London

For information address Bloomsbury Press, 175 Fifth Avenue, New York, NY 10010.

Published by Bloomsbury Press, New York

All papers used by Bloomsbury Press are natural, recyclable products made from wood grown in well-managed forests. The manufacturing processes conform to the environmental regulations of the country of origin.

LIBRARY OF CONGRESS CATALOGING-IN-PUBLICATION DATA

Kaplan, Michael, 1959–
Bozo sapiens : why to err is human / Michael Kaplan and Ellen Kaplan.—1st U.S. ed.
p. cm.
ISBN-13: 978-1-59691-400-1 (alk. paper)
ISBN-10: 1-59691-400-9 (alk. paper)
1. Errors. 2. Thought and thinking. 3. Practical reason.
I. Kaplan, Ellen, 1936– II. Title.

BF323 . E7K37 2009
153.4'2—dc22
2008044407

First U.S. Edition 2009

1 3 5 7 9 10 8 6 4 2

Designed by Sara Stemen
Typeset by Westchester Book Group
Printed in the United States of America by Quebecor World Fairfield

To both our families,
who forgive us our errors
and make none themselves.

Contents

CHAPTER I

From the Logbook of the Ship of Fools

Truth has an uncorrupted kingly bloodline; yet our world seems peopled with Error's bastards. Wrong thinking, reasoning that could never stand up to scrutiny, is universal and nearly constant. Why? Merely doubting is not a sufficient test to drive out error; nor is the classical machinery of formal logic. The Baconian revolution, however, established the scientific method and gave us a way to put ideas—in any language, at any scale—through the test of truthfulness. But this is a method we consistently fail to use in daily life, not just because it's a troublesome yoke, but because we don't naturally think that way. So is it more natural to be wrong? We'll see.

Stupidity does not consist in being without ideas—that would be the sweet, blissful stupidity of animals, molluscs and the gods. Human stupidity consists in having lots of ideas, but stupid ones.

—HENRY DE MONTHERLANT, *Carnets*

A DEMOCRACY OF DUNCES

People—other people, that is—make such *stupid*, easily avoided mistakes and never seem to learn from them. Try as you might to support the "in apprehension how like a god" theory of humanity, you're struck almost immediately by some counterexample that puts "quintessence of dust" back in the top billing. We may have been made in the divine image, but it was on Saturday night at the end of a long, hard week.

This puts us humans in the unique position of being constantly disappointed in ourselves, expecting a higher standard of reasoning and behavior than we ever actually achieve. "Well, *duh*" has become an accepted term in debate, presuming simultaneously that the truth will be obvious and that everyone will miss it. We easily spot and gleefully point out the fatuities of our opponents—and wonder, in lonely midnight hours, whether we ourselves are any less absurd. Error is democratic and egalitarian: go scrutinize the opinions of even the best educated, and you will find them still largely a patchwork of hearsay, authority, prejudice, and self-accommodation; basic illogicalities prevail alike in the labs of MIT as in the stands at World Wrestling Entertainment. Such universal dopiness (or, to give it its traditional name, "vulgar error") is not just a matter of being mistaken about the unknown—through excusable ignorance of string theory, say, or counterpoint, or Kierkegaard; no, it's being bald-facedly wrong in familiar things we say and do every day. We shamelessly yield to impulse and invent reasons afterward. We impute motives to distant figures and events of which, despite the global wash of media, we really know almost nothing. We shift our grounds, making the same issue a matter of fact or of principle as it suits our local purpose ("I'm a true believer, so my beliefs must be true"). We allow others to impose on us with slippery rhetoric and bogus statistics ("all *real* Americans will support me"; "200 percent lower prices!"). We cower from difficult truths and cry after comforting illusions. And yet, astonishingly, here we still are—the masters of creation. For idiots, we have been remarkably successful: our grand entrances may start on a banana peel, our sweeping exits lead into a closet, but we are the stars of this show.

The problem, like most, goes back to Genesis. The Bible has human history begin with a blunder: coming to know, through a temporary lapse in divine discipline, the difference between right and wrong. That blunder was the parent of all subsequent faults, errors, mistakes, and gaffes—because to *know* is to be allowed choice; and to choose is to have the option of choosing badly, assuming falsely, and indulging in all manner of specious self-justification. Since the exile from Eden, Right and Wrong have remained our intimate companions, presiding over every exalted and trivial thing we do, from declaring war to guessing the answer on game shows. Error is something that we both casually expect and find alarming to the point of apocalyptic despair.

"As it happeneth to the fool, so it happeneth even to me; and why was I then more wise?" Solomon was neither first nor last to worry about this: throughout humankind's triumphant progress similarly grim prophets have reminded us that our basic senselessness (now compounded by vast power) may soon lead us over the precipice. Yet despite these constant warnings we can never be sure exactly which of our many errors is the basic one, the fault we ought to tackle preferentially: meat-eating during Lent or not feeding the hungry? Sloth or excessive energy consumption? Our fractured family or our divided society? It's not surprising that, when the English Protestants first made confession public, they also had to make it all-encompassingly general: "We have left undone those things which we ought to have done, and we have done those things which we ought not to have done, and there is no health in us." Well, it's a start, at least.

Our dilemma may simply be a matter of probabilities: the intrinsic difference in likelihood between the one right way and many wrong ones. The path of righteousness is straight and narrow, but error can wander all over the plain. On one hand, we have the *valid*, the *true*, and the *good*: desirable ends, but only three. On the other, we have legion: bilge, bunk, and bosh; FUBB, FUBAR, and SNAFU; hokum, hooey, and humbug; rimble-ramble, whiffle-waffle, and yawp.[1] Having set down our few commandments, we open myriad opportunities to screw them up.

This idea was given a sharp point by the Italian economist Carlo Cipolla in his essay "The Basic Laws of Human Stupidity." Cipolla observed that the bad is statistically more likely than the good.[2] Of the four categories of human—which he calls the *helpless*, the *intelligent*, the *bandit*, and the *stupid*—three are composed of people destined by character to cause harm to others, themselves, or both. Cipolla's further laws establish that there is a constant irreducible proportion of stupidity in any human group (he includes college professors and Nobel laureates); that an observer will always underestimate the amount of stupidity in circulation and its power to do harm; and that the stupid person is the most dangerous of all, both because he does not intend the actual results of his actions and because stupid deeds by definition produce no benefit for anyone. "Day after day, with unceasing monotony, one is harassed in one's activities by stupid individuals who appear suddenly and unexpectedly in the most inconvenient places and at the most improbable moments." The reader chuckles, but not overloudly—this is too familiar.

Error is pervasive: it seeps into thought, word, and deed. It is universal: there are no Happy Isles where humankind is free of it. And like all blemishes, it is more obvious in others than in oneself. No wonder then that there have been so many attempts throughout history to free us from it.

THE FAILURE OF FALLACIES

In a world of stupid beliefs, doubt is the beginning of wisdom: if you can see that your neighbor's ideas have no foundation, you can at least avoid going inside them, even if that leaves you no place to live yourself. The Han dynasty scholar Wang Chong (a familiar university-town figure—prickly and poor, haunter of tea shops and secondhand bookstalls) had a keen nose for nonsense, and found much to offend it in the China of his time. The imperial government, first to rule over the whole Middle Kingdom, was constantly battling barbarians without and rebels within; under this stress a centralizing instinct (always strong in Chinese history) had condensed government, culture, and religion into a single lump of doctrine, to be swallowed whole. Confucianism, once a humanistic search after the harmonious life, had shrunk to a state church with Confucius promoted to godhead. Taoism, once a spiritual quest for tranquility in the flux of existence, had coarsened into a species of alchemy, touting secret immortality powders and condemning future generations to the rigors of feng shui. Official thinking was absurd and self-contradictory; but it was official, so anyone who went against it would need to be willing to seem awkward—and willing to stay poor. This was a job for Wang Chong.

His *Lun Heng*, or *Critical Essays*, ran through the body of contemporary superstition and muzzy thinking in eighty-five righteously indignant chapters. He asked, for instance, why aren't ghosts naked? The *clothes* of the dead had no vital force that would permit them to return. If setting up an earthenware dragon attracts rain, why wasn't the dragon-obsessed Duke of She, whose palace was encrusted with the things, flooded out? Lao Tzu says we attain great age by banishing ambition, yet many ambitious men live to be a hundred, while plants, seemingly the least ambitious of living beings, perish in a season. In a culture obsessed with signs and portents, Wang Chong's unwelcome message was the basic indifference of a world on which we live "as lice do on the human body." It's pointless inventing supernatural powers and beings: once you admit that at least some things just *happen*, you have lost any sure way to distinguish divine will from pure chance. Good and bad things occur everywhere in all ages. Why, then, is your good fortune a reward from God while your neighbor was merely lucky? When, say, Pat Robertson claimed to be able to steer hurricanes away from his broadcasting studios through the power of prayer, the Wang Chong question would be, "What about the *other* lives and property destroyed by hurricanes? Did these all belong to people who failed the prayer test?"

But even as he battled accepted foolishness, Wang Chong himself suffered from a fatal deficiency: he had nothing to put in its place. Some concepts might *seem* less vulnerable to doubt than others, but he had no uniform standards by which to test them. He was not even able to separate the two essential types of doubt: doubt from inconsistency and doubt from improbability—that is, things that don't fit what you've said versus things that don't fit what I've seen. What he needed was some kind of philosopher's stone to find the sense within nonsense, to tell meaning from meaninglessness.

Had he found the works of Aristotle on a bookseller's stall, he would have been able to take this next step. Aristotle holds a solar position in the history of thought: he is the source of illumination for so many subjects and the gravitational center around which so much later work revolved. His genius was *method,* sorting the richness of experience into logical categories and ordering these categories into chains of causality. The habit your science teacher insisted on, of defining your terms and specifying the relation between those terms before you went on, is a legacy from Aristotle, providing not just a powerful educational tool but also the means to isolate the valid from the fallacious in reasoning and speech—means we still use today in every lecture hall, courtroom, and debating chamber.

Aristotle's own teacher, Plato, had lacked such a method. He would ask, through the literary medium of Socrates, questions of the form "what actually *is* . . . ?" What is Virtue? What is the Good? What is the true meaning of these big concepts we all bring so easily and so unthinkingly into our conversation? It was no good saying, "Well, Themistocles is good; Aristides is virtuous." Examples are not explanations, any more than the scrawled diagram on the blackboard really is the proof in geometry. Plato's disputants constantly came up against this problem, because the most interesting ideas usually resonate beyond any explanation. Words indicate things they cannot contain. Often, the dialogues trail off into a state that contemporary rhetoric called *aporia*—the realization that there is no more that can be said. Not, for any Greek, a happy ending.

Plato's response to this embarrassing lack of a reliable clincher in argument was to posit, somewhere outside human existence, a world of Forms: perfect originals of which everything we see is a flawed copy. Forms relate to each other only one way—the right way, which we could confidently call Truth . . . if only we knew it. We come into this life having already known the Forms, so our ability to assign abstract qualities to things is a kind of *remembering,* a fleeting connection to the ideal knowledge we once held, shining and complete, in our bodiless minds. Stupid or unthinking people have simply forgotten more, and so are best governed by means of "noble lies," while people who love truth are more fit to rule as philosopher-kings, because they are more aware of what they have forgotten. All err, but at least the *aristoi* know they do. Well, yes—but determining who genuinely remembers most about the Forms is not a straightforward business: even Socrates' closest companions regularly found themselves at odds. What was really needed was a test that anyone, regardless of his degree of forgetfulness, could use to decide whether a given statement is valid or not.

Aristotle came up with the answer, and his solution was the one that works so often in mathematics: he turned the problem on its head. He reasoned, not from Form to example, but from example to form, from the world's things (nouns) to its qualities (adjectives). He left aside the absolute meaning of terms to concentrate on their use. Does this adjective *red* properly apply to this noun *chair*? Can it extend to include other nouns—*table*, *flag*, *China*? Does it encompass or nest within some other adjective—*scarlet*, say, or *colored*? If you take this noun-adjective relationship and then add four functional connectors—*all*, *some*, *not all*, *none*—and tie them up with the conclusive *therefore*, you will have all the tools you need to do formal deductive logic, the method we still use to decide whether a statement is *valid*; that is, whether it is consistent with a previous statement or line of argument.

It's a powerful technique. If I claim "*all* animals move," or "*none* are immortal," you can explode either of those statements forever with a single counterexample. Or say I propose the following syllogism:

1. All terrorists are extremists.
2. Omar is an extremist.
3. *Therefore* Omar is a terrorist.

You can expose the flaw in my reasoning by, for instance, pointing out that it is formally identical to "all chimpanzees eat bananas; my brother eats bananas; *therefore . . .*" Assuming you had read your Aristotle, you could casually mention that I had just committed the Fallacy of the Undistributed Middle, and that it is one of around twenty similar dodges (including the Illicit Minor, the Masked Man fallacy, and even the Fallacy fallacy) by which devious or ignorant wheedlers try to get you to agree that the world is divided up in ways it isn't or that properties of one thing can be transferred to another, when they can't. Such fallacies are the code violations of formal logic: once you spot one, you have condemned the whole jerry-built argument. The structure is clearly unfit for use; your opponent will have to take it down and construct another.

Aristotle was, in fact, too experienced a man to expect real people to actually argue using formal logic. He knew that speakers rely as much on appearance as on substance, just as "physically some people are in a vigorous condition, while others merely seem to be so by blowing and rigging themselves out as the tribesmen do their victims for sacrifice."[3] So to help us tell the mental athlete from the mere blowhard, he wrote *On Sophistical Refutations*, a handbook of rhetorical fakery as applicable now as it was in the fourth century B.C. It lists in order the various verbal equivalents of bustles, toupees, and elevator shoes that sophists use to tart up their unappealing doctrines. It covers question-begging, weak analogies, false generalizations, ad hominem arguments, appeals to force—all the slippery faults that, in logical terms, are not even wrong. In *Sophistical Refutations*, we have a catalog of every type of evasive maneuver, from amphibology ("I am opposed to war which dishonors our country"—comma, or no comma?) to tu quoque ("who are *you* to tell me drinking is harmful? You're a lush."). When you sense that some slick demagogue knows every trick in the book . . . this is the book.

These so-called informal fallacies are all the more powerful because they automatically engage the emotions. Logic's formal relations can be expressed in symbols ($\cap$, $\in$, $\not\subset$, and all that), but we have to define its terms in *words*—and words never simply mean what it says on the first line of the dictionary entry. Like subatomic particles, every word in our language has its characteristic combination of mass, charge, and spin. Do you talk about *workers* or *employees*? *The people* or *the mob*? *Faithful* or *credulous*? *Devout* or *fanatical*? Consider "liberal," which has been used in recent times to mean anything from "free-market economist" to "godless and gutless traitor."[4] Even carefully neutral words add a tone of unnatural blandness: when you try to sound calmly logical, you end up talking like Mr. Spock.

Words easily acquire connotations so emotive or politically charged that they become untrustworthy placeholders in any would-be factual statement; informal fallacies can then seize and use them to throw logic's equations out of balance, especially in the classic *loaded question*. "Are you saying that Wal-Mart's policy of providing low-cost products to lower-income communities is a bad thing?"[5] "What message does it send the terrorists if we cut and run?" "Do you refuse to take the military option off the table?" Have you stopped beating your wife? Moreover, connotations can attach to speakers just as much as to words: a sufficiently powerful personality or reputation can make the doubter hesitate to point out any gaping holes in the reasoning. So, concur because I'm bigger than you. Agree with my argument because I'm better educated, better groomed, or higher-ranking; or because I, my people, or the people I claim to represent have suffered a great wrong. Believe me because I have the skill to make my opponent look clumsy or foolish, or to distort his argument into something obviously absurd before attacking it. Accept my views on gun control because you liked my movies. And eat more meat: Hitler was a vegetarian.

None of this would have surprised Aristotle. His inventory of slipshod reasoning is depressing because it is both so ancient and so complete: in all our intervening history, we have not even learned new ways of being wrong. In fact, there are only two additions to this canon of error that we can call genuinely modern. The first is the statement with *no meaning at all*—the kind you find in French textual criticism:

> Singularities possess a process of auto-unification, always mobile and displaced to the extent that a paradoxical element traverses the series and makes them resonate.[6]

Or in business meetings:

> Going forward, we proactively accelerate our balance sheet through bottom-up empowerment in strategic partnering alliances.[7]

In both cases, the authors presumably had some vague "truth" in mind: there was something they meant to say but only they know what it was.

The second innovation, well defined by the philosopher Harry Frankfurt, is *bullshit*—that is, a statement whose author has no interest in whether it's true or not, nor even whether it is taken to be true.[8] It's just a warm pile of words, used to soothe opposition, enhance the speaker's image, or point out his allegiance to some generic good cause—offering hearers the emotional component of thought without troublesome meaning; verbal junk food, providing satisfaction without nourishment.

Why should we put up with all this? Logical fallacies, formal or informal, are not "legitimate differences of opinion" or "varieties of discourse"; they are *scams*: bogus, phony—falsehood strutting in the feathers of truth. Is there some particular reason we continue to fall for them? Well, it could be we're lazy or overtrusting; we need talk to fill time; or we simply prefer a vivid lie to a dull or awkward truth. We certainly share the weakness Wang Chong pointed out: we are blind to the random. We can't fully accept that unusual things can just *happen* in this world we crawl over—that "normal" is an artificial construct, a flattening for ease of calculation of the bumpy landscape of random events. When we find ourselves stranded on one of fortune's peaks or lost in a hollow, we stubbornly refuse to believe ourselves subject to mere occurrence. We are sure there must be a reason and will believe the man who provides one, even if—*especially* if—it is too far-fetched for us to test by local knowledge: UFOs, or the Bilderberg Group, or the Freemasons. Just as people would prefer to have succeeded through skill rather than luck, they would prefer to be the victims of conspiracy rather than chance.

There's also a deeper problem, though, that undermines deductive logic itself—a problem that seems at first glance no more than a puzzle for thesaurus writers: the distinction between the *valid* and the *true*. Aristotle gave us rules for the function of words in argument, but he didn't give us a means to attach those words securely to the real world we live in. For example, take this Dada-ish syllogism:

1. All Xqr are spoo.
2. 2VX7 is Xqr.
3. *Therefore* 2VX7 is spoo.

It is as *valid* a piece of reasoning as any, despite having no true meaning. It obeys faithfully the etiquette of formal logic—that is, the abstract relations of sets under operations. What we put into those sets, though—things real, things believed, things imaginary; caroming atoms or dancing angels—remains up to us. This means that speaking *logically* is no more a guarantee of sense than is speaking with a deep voice through a long beard.

BACON'S LEAP

Deductive logic is therefore not the gold standard for exposing error: it can reveal when we are being inconsistent, but not when our beliefs are nonsense in the first place. Luckily for us moderns, though, we do in fact possess (though only rarely use) a means to ensure that every statement we make about the world is as certain as secular matters can be: the *scientific method*. Almost exactly four hundred years ago, it leaped full-armed from the brow of Sir Francis Bacon to drive the legions of error off, one would expect, to grumble forever in some dark corner.

When Bacon was a little boy, he was too delicate to be sent to school and have Aristotle beaten into him with the master's stick. His clever mother instead gave him the run of the long gallery at Gorhambury and let him come to his own conclusions. The gallery was a renaissance prince's study palace, built in small. No opportunity for instructive ornament was wasted. Between the windows, portraits of great men; below, a few allegorical pieces and landscapes of distant countries; above, Latin quotations appropriate to each picture's subject; within, stained glass panels showing the flora and fauna of the four known continents; without, the real world—gardened, farmed, and wild. The gallery was an ideal counterweight to a library: a place where image, word, and fact supplemented but did not replace each other; where an observant mind could inhabit and reconcile all three. It shaped Bacon's view of the connectedness of understanding; that, like plants in a garden, senses needed to be trained up by study, speculations pruned by experience. He rejected the Aristotelian tradition of deductive logic as a "hunt more after words than matter."[9] Since valid deductive syllogisms could be constructed out of imagined terms, the thinkers who relied on them were like "spiders, who drew their webs out of their own entrails."[10] A true scientist should instead imitate the bee: go out, search, gather, compare, refine. Only by open observation of the world and isolation of its operations by repeated and combined *experiments* (a word of Bacon's coining) can we purge our minds of false preconceptions and vain imaginings, the "idols" that distract us from a true knowledge of nature.

Bacon's experimental science is inductive: instead of the broad *all* or *none* of deductive logic, it reasons from *this* and *that*—the individual phenomena of experience—compiling instances before making hypotheses, then contriving experiments in an attempt to prove hypotheses wrong. Any statement is *true* only to the degree that we have taken pains to disprove it but have so far failed in the attempt. This approach transformed people's idea of nature: before Bacon, the visible world was considered to be a feat of celestial handiwork, something to marvel at without inquiring too closely into trade secrets. After, it was a great mart of facts with which the alert mind could deal like a merchant, provisionally balancing doubt and trust in a series of bargains—that is, experiments. The test of truth moved from knowledge to experience; it was no longer a matter of finding confirmation in a book but of applying logic to repeated tests. Finally, we had a way to be *right* about things: any idea not supported by ordered observation could safely be scrapped, even if the whole cathedral library affirmed it.

One day in the late winter of 1626, the aging Bacon rode in his coach through London's snowy outskirts. As usual, his mind was traveling a separate course: the snow looked so pure and unblemished, like the best sea salt. Salt preserves meat from spoiling; might snow do the same? An Aristotelian would reason it out through definition and etymology; instead, Bacon stopped the coach, bought a chicken from a neighboring farm, killed it, and stuffed it with snow, testing his hypothesis—and bringing on a chill that killed him two weeks later. Many before him had died for their beliefs; he was the first to die for an experiment.

Gorhambury soon fell into ruin, but Bacon's new foundation for knowledge remained strong and secure. In the forty years after his death every modern science gained its basic principles—proving that what had previously been lacking was neither intelligence nor evidence, but a *method for being correct*. The test for truth by induction allowed discovery to become a shared enterprise; from Cádiz to Königsberg scholars built on each other's inquiries, or lobbed long-distance counterexamples into stoutly defended hypotheses. By the middle of the eighteenth century, most European thinkers simply *assumed* that this was the correct way to proceed, gradually revealing Nature's harmonious system in a spirit of fraternity. All else would be opinion, prejudice, bigotry, tyrannical self-interest—and thus unnatural and inhuman.

Everywhere knowledge seemed to be winning out over mere

opinion: when the Marquise de Pompadour found herself embarrassed at not knowing how her rouge was made, she nagged Louis XV into reversing his censorship of the *Encyclopédie*, because widespread information was clearly more important than state security.[11] Voltaire and his mistress, keen to uncover the nature of fire for an essay competition, turned her husband's château into a pyrometric laboratory, burning perfect cubes cut from the estate's trees and weighing the ashes. Travelers were considered to have wasted their trips if they did not come back with new data: heights and distances, pressed flowers and stuffed birds. Everyone, not just "natural philosophers," had a scientific hobby, passion or vocation; in such an intellectual atmosphere it would be entirely unsurprising to find a portly and nearsighted Philadelphia printer out in a thunderstorm with his kite. It was a time, in Immanuel Kant's phrase, of "man's emergence from his self-imposed infancy."[12] Kant and all his kind were convinced that, just as Newton's discoveries had illuminated astronomy, the simple abandonment of dogma and insistence on freedom to observe would bring Enlightenment—being *right*—to everyone. The formula was simple: *Sapere aude!* Dare to know!

STUMBLING DOWN THE ROAD TO ENLIGHTENMENT

What's happened since then? How far have we gotten with Kant's project? Science has amply fulfilled his hopes, but we ourselves—even after dozens of revolutions, thousands of new dawns—still fall a little short of enlightenment. The "scientific" social projects of the nineteenth and twentieth centuries, from Comtean phalansteries to state communism, have not delivered on their ideals. Our beliefs are still as confused, passionate, and childish as Bacon found them:

> What a man had rather were true he more readily believes . . . he rejects difficult things from impatience of research; sober things, because they narrow hope; the deeper things of nature, from

> superstition; the light of experience, from arrogance and pride; things not commonly believed, out of deference to the opinion of the vulgar.[13]

It's not just that some people believe in ESP, divine creation of fossils, or the power of pyramids to sharpen razor blades. It's not even that physicists crash their cars, Nobel Prize–winning economists tank in the markets, or moral philosophers cheat at tennis. Science itself cannot live up to its own standards. Although we publish and review the way Bacon said we should, we simply don't discover the way Bacon assumed we would. Our *inspirations* remain intuitive—rigor only makes its appearance at the write-up stage. This all-too-human habit can divert the wider currents of scientific thought: as Thomas Kuhn pointed out, shared paradigms tend to persist until they collapse under the weight of repeated counterexamples, rather than adapting continuously to accommodate new results. Prejudice also shelters in the private soul of each researcher.

The statistician William Cochran complained that scientists always came to him saying, "I want to do an experiment *to show that* . . ."—a phrase that would have made Bacon shudder.[14] It's like saying, "We'll have a trial *to prove him guilty*." Just as the ideal judge presumes innocence (because there are so many more ways to be innocent than guilty), the ideal scientist should *presume* that any given explanation is likely to be false—because there are so many more ways for randomness to act than can strict causation. It feels inhuman, though, to preserve such strict impartiality: even the most disciplined seeker after certainty maintains, in secret, his menagerie of pet theories.* If scientific method is the most natural way to avoid error, most of us remain strikingly unnatural—so no wonder we're error-prone.

---

* A typical example is the Swiss-American naturalist Louis Agassiz, who was willing to be accused of having "no *scientific* explanation" of the distribution of plant species around the world because, although his observations were careful and accurate, they would have shaken his prior theory of simultaneous divine creation. So, as the limb got longer, he just went further out on it.

THE JOURNEY AHEAD

The psychiatrist R. D. Laing said of the unconscious mind, "Perhaps something has lost touch with us."[15] Consciousness brings with it a gnawing sense of exile from the world of simple certainties. Our species is unique in noticing the poor fit between what we *know* is true and what we *feel* is true, between the logical and the intuitive. Other animals rarely trip (which is perhaps why it's funny when a cat does, and then pretends it meant to); they exist in a way that suits their circumstances. Goslings might take a naturalist's boots for their mother, macaques may have chaotic sex lives, but that's instinct, not foolishness. Is it *instinctive* for people—our doltish enemies, our spontaneous selves—to get things wrong?

This one question spawns many more: why is it, now that we have taken over the world, that we are apparently unable to stop ourselves from wrecking it? Why do we let belief blind us to evidence? Why do we abuse—and kill—each other in the name of unprovable abstractions? Why, despite having information that a Rothschild would envy, are our economic decisions so impulsive and haphazard? Why do we let celebrities tell us what to do? Why do we find mysteries compelling, and their solutions disappointing? And what has all this to do with the fact that all people speak the same way to babies and all cultures imagine heaven as flowery fields, laced with streams and studded with wide-spaced trees? These are some of the problems we will be taking up in the next six chapters.

Even if we don't think scientifically, we will need science to help us in our task: the very method we fail to apply to experience can turn its lens on us, examining and testing our own quirks of perception and decision-making and offering hypotheses to explain them. The emerging fields of behavioral economics, cognitive studies, neuropsychology, psychopharmacology, and evolutionary psychology have been examining the nature of human fallibility from wide but convergent angles. Researchers are beginning to identify the elements of our reasoning that consistently evade reason, to discover our biases where logic would expect balance. We are beginning to construct formal models that imitate, not how we *ought* to think, but how we do—and we can increasingly test how these models correspond to the structures and procedures of the living brain. And finally, at least in the vital matters of breeding and survival, we are coming to understand how some of our illusions may actually be useful: in offering an evolutionary advantage.

This book looks at the progress we have made in the science of human error. It traces the pedigree of the many pet theories we use to explain the world and explains why disproving them through logic generally has so little effect. It shows how difficult we would find living up to our own standards of perfection and explores the hypothesis that our cognitive and logical failures may actually be the fair price we pay for our extraordinary success as a species.

There will be no self-flagellation here; if anything, the message should be hopeful: if we are more conscious of the shared mistakes that define our humanity, we can live more comfortably in this world we have created but will never fully understand. It will not make us right, but could help us be wrong better.

CHAPTER II

Idols of the Marketplace

The desire to squeeze error out of practical life has given us economics, an admirable if dour discipline that takes the multifarious concept of Worth and converts it into a measurable and universally convertible substance: Value. Looking at exchanges of value makes clear how our arithmetically challenged ancestors regularly fell into error, but it also reveals the myriad ways we ourselves fail to recognize equivalences or trade-offs. The point is not that we have paid too little attention to economists but that Worth may really be as we once thought it was: unique to certain times and places, to be sought for itself, not its price.

Riches are a good handmaid, but the worst mistress.
—FRANCIS BACON, "Of Riches," *Essays,* 1597

ONE of the disadvantages of being a fish is the inability to scratch: having no hands makes it difficult to rid oneself of loose scales, patches of thickened skin, and the minuscule crustaceans that take up unwelcome residence in one's gills and mouth. Even for a fish, habitually so cool and undemonstrative, it must be maddening.

This predicament explains a remarkable phenomenon: marine cleaning services. Small, highly visible fish—the best known being the blue tropical wrasse *Labroides dimidiatus*—set up shop in an open section of a reef, where larger fish of other species come and wait with mouths agape to have their blemishes and parasites nibbled away. It's a mutually beneficial arrangement, despite some natural tensions: the cleaner fish can be tempted to cheat, taking a surreptitious nip from their customers; or a too-hungry visitor might suddenly decide to treat this salon as a restaurant. That the arrangement persists means it must be the result of some larger calculation of worth, which brings it into the realm of economics.

Redouan Bshary of the University of Neuchâtel in Switzerland has watched cleaner fish for over a decade and his findings reveal a marketplace of considerable sophistication.[1] Each cleaner has a small territory; its customers include a mix of local residents and wider-ranging fish that have a choice of where and when to stop for a quick spruce-up. Any such "choosy" fish, like a tourist in the bazaar, will get plenty of attention from the cleaners. It will usually watch to see how each cleaner works on its existing customers, and if satisfied by the apparent bustle and dedication, pose to show it is ready for grooming.[2] When cheated, it tends simply to move on, metaphorically shaking the dust from its fins.

Local fish build a closer relationship with their cleaners before the actual service begins ("How's business, Joe?" "Eh, so-so"), but often find they are made to wait when a choosy fish jumps the line ("Step right in, sir!"). In this respect, cleaner fish act the way banks do, pampering new customers because they know existing account holders will usually put up with poor service. If cheated, though, the locals don't simply move on; they make a fuss, punishing the cleaner. Cleaners, for their part, gauge finely whom they can cheat and wisely play fair with major predators: nobody diddles the Don.

The Art of the Deal has here been mastered by creatures with brains you could fit on a fingernail. If you consider rationality to mean acting in a way that secures a long-term personal advantage—what the ever-poetic economists call *maximizing utility*—then we are surrounded by other rational animals. Thrifty scrub jays not only remember where they have cached food; they shift their hoards around if they think they have been observed (the number of moves also depends on the attractiveness of the morsel: worms, say the researchers, are the Belgian chocolates of the jay world).[3] Mother baboons demand grooming services from those who want to hold their new babies. When there are fewer babies around, the amount of grooming required goes up, suggesting that the hand that rocks the cradle is also the Invisible Hand of the market.[4]

In the Glimcher lab at New York University, macaque monkeys peer into computer screens, shifting their attention to maximize their expected utility (in their case, earning sips of fruit juice) with the intentness and efficiency of their human cousins farther downtown on Wall Street.[5] In the Chen lab at Yale, capuchin monkeys use tokens to buy and trade treats, allocating budgets to get the most of what they like best.[6] Life is choice: impulse tempered by prediction, effort prompted by reward. All around us, animals are making these choices—and making them pretty well.

Why, then, do we humans so mess up our finances? Why do we sell the stocks that make us money and hang on to the losers? Why do we scrimp on small purchases and waste hundreds of dollars on large ones? Why do we keep balances on our credit cards when we have cash in the bank? And why do none of us save enough for old age? We buy lottery tickets but spurn insurance, fill our attics and garages with fast-depreciating junk, and pursue happiness through the one means—money—that we already know cannot buy it. Americans may think more often about money than about sex (as one survey reports) but clearly to less purpose—or there would be many fewer Americans.[7]

THE DREAM OF RATIONALITY

It was not supposed to be like this. In the hopeful eighteenth century, educated people assumed that the free play of individual interest would bring reason to financial affairs. Clearing out medieval irrationality—France had six different tax regimes for salt—would open the way for comparative advantage and make trade, like good conversation, an exchange of superiorities: your champagne, truffles, and foie gras for my coal, cottons, and cheap tin trays. Rational agents, pursuing their own ends, would automatically contribute to the public good by providing the products and services others valued enough to buy—an arrangement most clearly expressed by Adam Smith in a famous passage from his *Wealth of Nations*:

> It is not from the benevolence of the butcher, the brewer, or the baker that we expect our dinner, but from their regard to their own interest. We address ourselves, not to their humanity, but to their self-love, and never talk to them of our own necessities, but of their advantages.[8]

The last element necessary for rational economics was the *utility function*; and this was supplied by the mathematician Daniel Bernoulli in 1738, when he realized that money, like all goods, means different things in different quantities, just as the first bite of bread means more to the famished laborer than the last wafer-thin mint to the bloated gourmand. Value is not a linear function, nor is money the only value. Our rational economic choice, therefore, should be what secures the highest expected level of what we personally value most. That, certainly, is what the monkeys seem to be doing.

Human lives, of course, involve many more variables than juice or no juice. Anyone who has known, or been, a small boy will be intimately aware of the breadth of the concept of "good stuff": trading cards, bottle caps, neat packaging, particular sizes of pebble or stick—all are objects of great intrinsic *worth*, to be accumulated indefinitely, whose disposal by thoughtless mothers prompts howls of justified reproach. There is something of the little boy in every age and sex: most women have at least one pair of shoes they treasure too much to wear, most men a power tool that is far too potentially useful to put away. Gold may be a primitive measure of wealth, but we still thrill at the thought of all those ingots stacked in the vaults of Fort Knox. Credit rules the real economy, but fiction requires a briefcase full of cash. Value in classical economics is supposed to be *fungible*—that is, it should transfer without a shudder from one valued thing to another. But imagine paying for your half-million-dollar house, not with a banker's draft, but with bills (counting out, one by one, five thousand portraits of Franklin), or with fifty pounds of gold, or the unpaid servitude of your two teenaged children for the next ten years. It's the same dollar *value*, but somehow they don't feel equivalent.[9] Switching from the concept of Worth to Value may be a leap necessary to reach the promised land of rational economics, but it is a leap we find difficult to make.

Nor is it the only one; we must also learn to put a price on the uncertainties of life. Expected utility multiplies value by probability: economic theory assumes that we make our choices knowing, or at least being able to guess, how *likely* are the events that offer the various payouts. This may not sound entirely like real life (would you care to guess, say, where the Dow will be on the day you retire?), but perhaps it's an assumption worth making for the sake of the model: at least in the abstract, having clear probabilities should help us make rational choices. So here is a nice abstract challenge devised by the Nobel Prize–winning economist Maurice Allais in 1953. See what you think.

His first question offers you a choice: would you prefer A) a guaranteed million dollars; or B) a gamble: an 89 percent chance of getting a million dollars, a 10 percent chance of getting five million, and a 1 percent chance of crapping out and getting nothing at all?

Once you've made your choice, consider a second, separate proposal: would you prefer C) an 11 percent chance of getting a million dollars, with an 89 percent chance of getting nothing; or D) a 10 percent chance at *five* million dollars, with a 90 percent chance of getting nothing?

If you didn't choose A and D, you're not human—you're a computer. A computer would say: "A offers a clearly defined chance of winning a million dollars (in this case, certainty); so does C. B offers a 10 percent chance of winning five million dollars; so does D. Therefore, if I prefer A, I should prefer C; if I like B, I should like D. I could not prefer A and D at the same time; that would not be logical." Yet that's exactly what we do; we go for the sure thing when we're offered it but the gamble when we might otherwise lose out. Even L. J. Savage, who *invented* subjective expected utility theory, preferred A and D until he "corrected an error" in his instincts.[10]

FOLIES DE RICHESSE

Most people's instincts remain uncorrected: our inability to transfer subjective value from one good to another or to assess consistently the probabilities of future events goes some way toward explaining the extremes of irrational money behavior that appear in every time and culture. We all know tales of wastrels who blew huge fortunes on dice, horses, or cocaine, or of crabbed pinchpennies who lived in squalor, gnawing crusts, with rolls of banknotes stuffed in their dusty teapots. But there are subtler forms of these madnesses that may be even more instructive.

In 1782, the twenty-two-year-old William Beckford wrote his romance *The History of the Caliph Vathek* in a single sitting, as a kind of spiritual self-portrait:

> He surpassed in magnificence all his predecessors. The palace of Alkoremmi, which his father Motassem had erected on the hill of Pied Horses, and which commanded the whole city of Samarah, was in his idea far too scanty; he added therefore five wings, or rather other palaces, which he destined for the particular gratification of each of his senses.[11]

Born to an immense fortune, trained in all the arts by the masters of the day (Mozart, for instance, was his piano teacher), Beckford knew he had the means to accomplish anything his flamboyant imagination suggested. He toured Europe restlessly, buying, buying, buying: paintings, sculpture, furniture, books—everything that might gratify the senses. Unlike many compulsive collectors, he bought well: a Beckford provenance is still highly desirable. He returned at last to England and his country estate at Fonthill, knocking down his father's huge and comfortable house and setting about creating a suitable home for his collection:

> The palace named "The Delight of the Eyes, or the Support of Memory," was one entire enchantment. Rarities collected from every corner of the earth were there found in such profusion as to dazzle and confound, but for the order in which they were arranged.[12]

Fonthill Abbey, designed in a style we might now call Hogwarts Gothic, was vast, at once self-indulgent and uncomfortable, luxurious and jerry-built. Beckford was impatient of such petty constraints as structural studies or architects' plans. He knew what he wanted and knew he could get it right now. All that was necessary was money—money he had in bucketsful. The three-hundred-foot tower that would dominate the building fell down twice during construction, but Beckford simply recruited every available workman, bribing them to labor day and night by providing limitless food and beer, and therefore had his house finished when he wanted it—although much of it had been built in the dark by men who were very, very drunk.

Fonthill failed in its role as the Palace of Delight of the Eyes. It was gloomy and cold. Its football-field-long vistas, lined with perfect pictures, seemed only to confirm the isolation of its master. Beckford sat alone at a table set for twelve. He disliked the bell system for summoning servants, so he stationed a man in each room to pass along his wishes. Thus the only sign of life in the great palace was the echoed whisper of "cheese" or "fruit" drifting down the lonely corridors toward the pantry.

In the end, having run through an inheritance roughly equivalent to six hundred million dollars today, Beckford sold the abbey and most of its treasures, retiring to a house considerably less grand than the one he had so confidently demolished.[13] He had attempted, with skill and determination, to turn cash into happiness—and failed. Three years after his ruin, his monstrous tower collapsed for the last time: completely—to powder—and silently.

> After awhile I never went out. My friends came to see me and I received them in my rooms, but at night I went up to the fourth floor and slipped into a storeroom. There, under a bed on top of which was piled furniture and rugs, I slept. For days I did not leave my room and lived on crackers and raw eggs, and all the time those "schemers" were trying to get my money.[14]

The coin has two faces. Turn it round from the gilded profile of Beckford and you reveal the plain, jut-jawed visage of Hetty Green, "the Witch of Wall Street." Born in 1834 to a successful New Bedford, Massachusetts, whaling family, she learned the money trade by reading out the financial news to her father and grandfather as they worked at their desks. After their deaths, she managed with great difficulty to secure control of her inheritance of one million dollars. In the course of the next half century, she would increase that to a hundred million (the equivalent of $1.7 billion today), making her the richest woman in the world.

Mrs. Green's great weapon was self-knowledge: "I don't believe much in stocks. I never buy industrials. Railroads and real estate are the things I like."[15] She lent money when it was tight (once even bailing out the city of New York) and swooped in to buy when others were caught up in panics. She was careful, well prepared, and persistent, letting the laws of interest do the rest: after all, a hundredfold increase in fifty years represents only a compound rate of less than 9.7 percent per annum.

To gain from interest, though, means maintaining capital; like an army in being, it gives you the power to dictate events. Mrs. Green took this maintenance to extremes. She kept no office, preferring to scrounge space from her bankers. She moved regularly between tenement flats in New York and Hoboken, New Jersey, in order to avoid paying residence tax in either state. She would wear the same black dress until it wore out and then replace it with an identical one, bolstering it in cold weather with judiciously stuffed newspapers. When her son broke his leg, she removed him from the hospital and kept him at home, where the leg developed gangrene and had to be amputated.

It may be that Mrs. Green's reputation for pathological stinginess stemmed from her isolated position as the only woman in the boys' club of Wall Street; it may be her closeness was a form of self-defense, protecting her private character as well as her capital. Whatever the truth, it's clear that she gained little pleasure from her fortune beyond its simple existence and continued growth. To spend, for her, was to lose; the pleasant and charitable purposes to which her descendants put their inheritance will not have gratified her ghost.

Just as Beckford and Green were not actually the monsters of traditional morality tales, the point of their stories goes beyond mere proverbial wisdom. These two, with all their exceptional intelligence and determination, proved unable to get money to perform the most basic thing money is designed, in theory, to do: make life enjoyable. Their fortunes were not fungible into other values. This implies that our own lesser irrationalities (the false belief that life will be transformed if we buy that handbag; the perverse thrill of saving rubber bands) have a deeper cause than the simple inability to think like an economist. There seems to be something fundamental about using money as the stand-in for worth that causes us to abandon common sense.

THIS IS YOUR BRAIN OUT SHOPPING

Economics is the science of what *ought to be* if calculations of value were rational. The last ten years have seen the appearance of an entirely new field, revealing what these calculations actually *are*: neuroeconomics, the discipline that looks at the physical processes of preference and decision in the brain. In universities across the world, people and other primates are making buying decisions while wearing caps of EEG contacts or lying in the doughnut holes of MRI scanners. The results, with new discoveries every week, are fascinating.

Let's start with value. What makes something *valuable*? Do we mean the same "good" when we talk about a "good bacon sandwich" as about a "good trumpet voluntary?" Well, perhaps so: Camillo Padoa-Schioppa and John Assad at Harvard have found individual neurons in the orbitofrontal cortex of the brain that encode the idea of value in economic choice, providing a common scale of preference that allows primates like us to decide between, say, apples and oranges without having to convert both into an intermediate good such as money.[16] Moreover, the preference encoded by these neurons is consistent and robust. The rhesus monkeys in their experiment would, for instance, habitually prefer one drop of grape juice to three drops of apple juice, but not to four. Experience does not necessarily reset the brain's preferences, which explains why the champagne of newfound wealth may never taste better than the beer of youthful poverty.

At the same time, neurons in the lateral intraparietal area of the cortex, an area of the brain believed to translate visual information into eye-movement commands, seem to activate based on the *expected* gain from looking at one thing or another.[17] This suggests that part of our brain may indeed operate according to subjective expected utility, calculating benefits and probabilities, but the calculation isn't necessarily conscious. We may believe we are making decisions, but our attention has been predirected by the financial adviser in our skulls.

The more researchers look, the more diffused through the brain value judgments appear to be.[18] We anticipate financial gains—in the absence of a choice—in the nucleus accumbens (a very old, very basic feel-good area of the brain), while we evaluate those gains, once made, in the medial prefrontal cortex. We anticipate physical pain and financial loss in the insula, described by some researchers as the area that governs the sense of visceral states—literally "gut feelings."[19] Having to make a choice sets these areas against each other: gauging a risky investment will see activity switching between insula and nucleus accumbens as we come to decide that the risk is worthwhile.

"Good stuff" seems to excite us in a different way from financial gain: the mere sight of it activates the same pleasure-related areas in the mesolimbic that are involved in its enjoyment (this explains the lasting appeal of paintings of luscious food, gleaming silver, rich fabrics, and bosomy nymphs—there have been no famous still lifes of a T-bill). In a recent experiment, men and women in functional MRI scanners were shown a range of desirable products (including, as it happens, Belgian chocolates—although not worms) at a range of prices and were asked to decide whether to buy at the given price.[20] The results showed a civil war in the brain: the preference and pleasure centers lit up at the sight of the good things, the pain center at the sight of an excessive price. Sufficient price pain would damp out product pleasure and the subject would not buy. None of the higher decision-making regions seemed to be involved, perhaps because the choice did not involve shifting attention, moving, or gaining new information—it was a simple matter of *yum!* versus *ouch!*

If, therefore, buying decisions keep us in tension between opposite urges, the rope in a constant tug-of-war, does our reason act as the wise referee, ensuring the two pulls balance? Traditional economics assumes so: it values "yum" and "ouch"—money in, money out—the same. But we clearly don't. The glow we might feel at adding another thousand dollars to our 401(k) account is nothing compared to the sinking sensation at discovering we accidentally sent off a hundred dollars to the dry cleaner's in a jacket pocket. Perhaps, though, this is a matter of ignorance: if we were economists, we might know better.

Or not. Paul Samuelson, one of the giants of neoclassical economics, had lunch with an MIT colleague in 1963. In the slightly challenging way of many academics, he proposed a gamble: put up a hundred dollars for the chance of winning two hundred dollars on the flip of a coin.[21] The colleague demurred, even though the odds were better than for many investments—but, being an economist, he said he would happily accept a *long sequence* of such bets. Samuelson was outraged: if something is a good bet in the long run, it's a good bet once; if it's too rich for your blood once, it should be a poor choice in the long run.

The colleague, though, was expressing a feeling we all share: yes, the probabilities are on my side, but if we did it only once, I might *lose*—and I wouldn't like that. Equivalent actions change their importance to us depending on the immediate, local risk: walking along a wide plank means different things when the plank is on the ground or twenty stories up. So if I offered you a bet that apparently worked to your advantage, you would doubtless consider not just the less-than-even chance of losing but also a host of other worrisome possibilities: *Will the game be run fairly? Will he really pay out if I win? Is there a catch I haven't noticed? What kind of mug does he think I am?* All these skew our estimate of acceptable odds—and all, significantly, would be gradually dispelled by repeated play. Samuelson's colleague had a point.

LOSS AVERSION AND OTHER UNSHAKABLE HABITS

Daniel Kahneman and Amos Tversky—whose study of these wrinkles in expected utility launched the burgeoning field of behavioral economics and gained Kahneman the Nobel Prize in 2002 (Tversky had, sadly, died before then)—managed to quantify this human unwillingness to take an even-money bet, which they called "loss aversion." All in all, their subjects would refuse to gamble unless the average expected payout from the game was at least *twice* the amount they put at risk—the kind of returns even a con artist would blush to advertise.[22]

Amazingly, Keith Chen's tufted capuchin monkeys at Yale seem to make nearly the same calculation: offered a choice between a sure portion of treats and the chance at more or nothing, they refuse to gamble unless there's at least two and a half times the usual portion on offer.[23]

The capuchins also revealed another quality familiar from human studies: *reference dependence*. That is, they valued things in relative but not absolute terms. If the experimenters showed the monkey a portion of apple pieces and then added some more, bringing the total to X, the monkey was delighted. If, however, the experimenters showed more than X and then *removed* pieces down to that total before handing them over, they had to deal with a very grumpy monkey. We primates hate to lose—and we gauge loss not just by what we already have but also by what we *could* have had. I could have bought Google at $150; I could have held on to the apartment in Boston; I could have kept my rock band together . . . the inner capuchin is never satisfied.

Reference dependence goes deeper than simple judgments of value: it's an essential aspect of how we perceive the world. The nineteenth-century German psychologist Ernst Weber gave his name to a law that links stimulus and perception asymmetrically: for us to *perceive* that something like loudness, brightness, or weight is increasing linearly in uniform steps (1 + 1 + 1 + 1), it actually has to increase geometrically (say, 1 × 3 × 3 × 3); that's the reason why things like star magnitudes and decibels are measured on a logarithmic scale. This wrong-end-of-the-telescope quality in perception makes small changes in the status quo seem big, while big changes, given a large enough context, seem small; so the general who lectures his college-bound son on thrifty living is likely to be comparatively unconcerned about the $200 million cost overrun in his $100 billion departmental budget.

True, that $200 million did not come out of his own pocket, while the allowance did—the reference included *self* in one instance and not the other—which points to another discovery by Kahneman and Tversky: what's called the *endowment effect*.[24] The first formal experiment revealing it worked like this: the professors distributed a number of college coffee mugs randomly through a class of economics students. These mugs usually sold at the bookstore for around $6; the students were told that if they wanted a college mug but did not have one, they could buy one from another student, assuming they could agree on a price. The expectation was that around half the mugs would change hands, but in fact hardly any did. Why? The prospective buyers would not offer more than $2.25 to $2.75 (after all, they knew that the mugs had been given out free), while the owners were unwilling to sell for less than $5.25. The mugs were *theirs*; even if they would have preferred the money, they were not going to part with something personal at a discount to its "true" value. A subsequent experiment showed that the mug owners didn't actually *like* what they had more than an equivalent alternative; they just hated to give it up: "a poor thing, but mine own." Once what you own actually does have value to you, the spread widens further: Dan Ariely, then at Duke University, attempted to set up a market in the rare and much-desired passes to Blue Devils basketball games, which are distributed to students by lottery. The average price at which winners were willing to part with their passes was $2,400; the average price losers were willing to pay was just $175. None changed hands.[25]

Endowment can operate in surprising ways, one of the best known of which is the lost-ticket problem.[26] Filled with pleasurable anticipation, you arrive at the theater box office. The tickets cost $10 (you can tell right away this is a thought experiment, not a real theater). As you come in, you discover A) that you had dropped a $10 bill when you parked—do you still buy a ticket?; or B) that you had dropped the ticket you'd previously bought—do you buy a second ticket? It turns out that 88 percent of subjects answered yes to A, but just 46 percent to B. Money is only potential—it hasn't yet really delivered its worth—but that was *my* ticket, I'm not going to pay for it *twice*.

Thus far the academic studies; but what do these hardwired, primate-brain habits—these unavoidable deviations from classical economics—actually make us do in the economic realm?

Are you, for instance, sheltering a dog in your portfolio? Most of us will acknowledge, if only to ourselves, that some of our investment decisions have been real stinkers; what's surprising, though, is how we cling to them. An exhaustive study by Terrance Odean, covering the trading records of more than ten thousand customers of a discount brokerage, revealed that people routinely sell their winners too soon and hold on to their losers like old friends, even though investment losses are tax-deductible.[27] The result is that, on average, the stocks people sold did 3.4 percent better over the next year than the ones they didn't. We don't just throw out the baby; we keep the bathwater.

What's the reason? Well, taking a loss is an admission of failure. It reduces the original holding—*my* money. But selling a winner brings in apparent gain—other people's money. There's also a touch of the *gambler's fallacy*: the false belief that the further away something is from its average value, the sooner it will snap back. In this case, we believe the "average" for an investment is what we paid for it. "I'm just holding it until I can break even"—but people who said that about their 1929 RCA stock would still be waiting.[28]

Will you sell me your house for fifty thousand less than you paid for it? *What, are you kidding?* Despite whatever is happening in the real estate pages, the amount we paid for it determines its worth; we'd no sooner part with property at a loss than Kahneman's students would part with their mugs. The shared desire to maintain and increase prices not only creates the more painful aspects of any bubble market (expensive, useless advice; hopeless indebtedness; irresponsible lending; sudden collapse) but also undermines the function of prices in general: to specify value. When the U.S. housing market first began to wobble, there were certain areas where prices apparently remained high, but where sellers offered to throw in new kitchens or cars with the deal. They weren't taking less, they were just, uh . . . being generous: that is, they were giving up worth to maintain value.

IRRATIONAL LACK OF EXUBERANCE

Look back for a moment at Allais' paradox (on page 23), specifically at the choice between A and B. A, the sure thing, is what most people choose, but B is actually a better bet over the long term. Our preference for predictability often means we choose less successful investments:

keeping cash in low-yielding bank accounts rather than buying a range of equities, which, though frighteningly unpredictable, do provide a better average return.

Loss aversion and reference dependence conspire to erode our savings. We smile at the regular increase on our bank statement, never noticing that at least 2 percent of that growth will be nullified by inflation. We frown at the occasional drops in share prices (and even if they rise over time, the more often you check them, the more times you will notice a relative drop), registering a pain that is double the pleasure of seeing a rise, and forgetting, in our distracted way, that we have also been paid dividends.

Those of us who do invest in the stock market can easily become victims of our loss aversion, spasmodically leaping in and out in our eagerness to catch the peaks and miss the troughs. Sadly, the old speculator's proverb that "time in the market beats timing the market" is absolutely true. A study by Nejat Seyhun of the University of Michigan showed that if you had remained invested in the U.S. market for the thirty years between 1963 and 1993, you would have seen the equivalent of an 11.83 percent annual average return (a figure Hetty Green would envy), despite the fact that this period included the miserable seventies.[29] If, however, you had traded actively and happened to miss the ninety best days in those thirty years—the equivalent of one summer in a generation—your return would only be 3.28 percent, without even taking transaction fees into account.[30] An intelligent, informed investor might be able to gauge whether the market is generally over- or undervalued, but for most of us, attempting to avoid local losses by active trading is rather like jumping on a trampoline to avoid the effect of earthquakes.

THE FLAVORS OF MONEY

The endowment effect compounds our financial confusion by making us unable to think even of money as just money: it's always money *from* somewhere and money *for* something. The money from an unexpected tax refund or lottery win rarely finds its way into the mental box marked "long-term savings"—it's carefree, bubbly money that demands at least a family dinner out. Money for old age or medical emergencies is untouchable—we would rather run up credit card debt, paying extortionate interest, than dip into such a sacred fund. Give people a cash bequest and they will invest it as they choose; let them inherit a stock portfolio and they will preserve it unchanged—even when it does not fit their investment style—as if it were the ancestral estate or the family silver.[31] And why do we have so many separate synonyms for financial outgoings: costs, fees, charges, commissions, options, packages, add-ons, payment structures, honoraria, sweeteners, and tips? Because we cannot see pure quantitative value when we consider separate qualitative goods.

No purchase is ever purely economic. Buy a new car, drive it off the lot, and listen for that jingling sound as it loses one sixth of its value in the first hundred miles.[32] Stick to your beloved old rattletrap, pouring in cash for increasingly serious repairs, until all you can sell it for is ten dollars and a promise the junkyard dog won't bite you.[33] In either case, you have been caught out by the endowment effect—because your car is never just a budget item; it is an expression of your inner character, from the free-ranging outdoorsman in his SUV to the Italian film diva in her white roadster.

We cannot isolate any purchase from the rest of our emotional landscape. If you feel generally disgusted with life, you become less willing to pay good money for anything, but you are also more willing to sell your own possessions for less. It all becomes mere *stuff*: the endowment effect disappears.[34] If you are sad, you willingly part with the material companions of your sadness, but will pay over the odds for something new (that might, after all, change your life); the endowment effect *reverses*. Remember, then: going shopping when you're low can be just as dangerous as going to the supermarket when you're hungry.

Big enough purchases can switch off anyone's sense of fair monetary value. Electronics salesmen know that once the customer has agreed to buy the various components, it's a cinch to sell the gold-plated cables to connect them and the veneered-chipboard cabinet to display them in. No sane family ever really *needed* a house with a two-story foyer, a cathedral ceiling, a triple garage, gateless gate piers, a granite breakfast island, a built-in vegetable steamer, dual vanities, and a dog shower.[35] But the real estate agents do: such up-front extras move a property quickly, making it what's called a "ten-minute" house. Only later will the buyers realize they have paid not for a place to live in but for an assemblage of selling points.

If all money comes with a mental label, the worst, of course, is "wasted." History provides so many examples of people who attempted to recoup the waste of some of their money (or their shareholders', or their taxpayers') by wasting far *more* that you'd think the lesson should have been learned by now. Once a government has built a white elephant like the Brabazon (a lumbering, fault-prone British transatlantic airliner of the late 1940s that took seven years to design, flew for only one hundred hours, and cost every man, woman, and child in the country five shillings), that should be it. But the same government later built the Concorde (which took fourteen years to design, flew one hundredth of its expected passengers, and cost every Briton ten pounds). Nor are governments solely to blame: the Convair jetliner lost General Dynamics more than any single product had ever lost a company; the CED video disc finally killed off RCA.[36] Thousand-percent cost overruns, sinking airports, roads to nowhere, pointless mergers, dead-end software projects—most flops reveal their essential hopelessness long before the funding tap is turned off, yet a combination of loss aversion, endowment effect, and sheer wishful thinking keep the good money flowing after the bad.[37] Folklore may point encouragingly to Edison's ten thousand failed experiments in lightbulb filament material; reason, though, agrees with W. C. Fields: "If at first you don't succeed, try, try again . . . then quit. No sense in being a damn fool about it."

We get caught in these traps because our problem with loss isn't so much its absolute value (although that can soon become the problem); it's the *being wrong*, making the chump's choice, exposing ourselves to the ridicule of the world. Buying a Betamax video recorder or buying an Edsel were actually entirely different kinds of failure (one was technically but not commercially superior, the other a four-hundred-million-dollar design disaster), yet they felt equally embarrassing.

The desire not to have made such a risibly poor choice can itself induce a kind of paralysis. Modern Germans suffer this in the form of the *Neuwagenverweigerungsproblem*: being unable to decide which automotive technology will prevail and therefore refusing to buy a new car of any kind; the average German automobile is now eight years old.[38] Banks are full of the cash of people who waited through the entire bull market to invest because they were afraid they might buy at the peak. And this, from personal experience: an old woman in Scotland, practically a pauper, willingly continued to pay 25 percent more for electricity than she needed to, simply because switching suppliers would have meant admitting (particularly, one suspects, to her daughter-in-law) that her previous choice was wrong. That would have been a loss far greater than mere money.

Not everyone has the same loss aversion. Some of us are extreme dreaders—the sort who, like lab mice, will choose to receive a stronger electric shock now just to get the anxiety over with—while others have less active imagination for pain.[39] These latter are the people who tend to make risk-seeking investment mistakes: gamblers. When making financial decisions, their bumptious nucleus accumbens trumps their timid insula every time.[40]

GAMBLING WITH YOUR INSULA

No one has ever made a successful movie with a hero who makes only sensible, cautious financial decisions. We are fascinated by gamblers, probably because most of us are so risk-averse; we mock scaredy-cats although, with few exceptions, we know that they are us. It's possibly a good thing, however, that we do not value caution too highly: in a world subject to sudden change, humanity needs at least a minority of risk-takers, just in case. This may be the reason most teenage girls have at least one boyfriend they should never, ever marry—and their fathers, as ex-teenagers, recognize that boyfriend with horror.

The dichotomy between swashbuckler and milquetoast may be overdone. Most of us do actually take risks, or we would never get out of bed; we let strangers fly us places and eat food in restaurants, prepared by unseen (and possibly unwashed) hands. Similarly, *successful* gamblers don't seek risk blindly: they calculate it. Chris "Jesus" Ferguson, poker champion, may look like someone out of a Willie Nelson song, but he sounds like someone from the RAND Corporation—all game theory and recursive Bayesian analysis—and both his parents are math professors. When gambling isn't a mug's game, it's no more thrilling than chess. The true risk-seekers are the ones who *lose*—those on whose wallets the whole city of Las Vegas was built. What sets them apart? Well, at the risk of sounding insulting, there's something wrong with their brains.[41] In extreme cases, damage to the prefrontal cortex can knock out the ability to relate risky judgments to an unpleasant future.[42] People who have suffered this kind of damage may still be able to do the math and tell, in the abstract, that something is a losing bet—they just don't really care about it. They lack, or ignore, the gut feeling that says when to fold 'em.

Less extremely, risk-seeking gamblers seem to have a reduced capacity to imagine future states, either bad or good.[43] Loss-averse people show heightened pain responses as potential losses grow, but also heightened pleasure at the prospect of big gains; risk-seekers have a deadened quality in both responses. They need big stimuli to feel anything—the financial equivalent of bungee jumping or Russian roulette. It's as if only by risking their whole futures can they achieve the same thrill the rest of us get by beating our kids at Go Fish.

Casinos aim to provide just such a thrill by surrounding players with the allure of success. The games offer a false sense of personal control, even if it's just pulling the handle of the one-armed bandit. The rooms are bright and shimmering yet isolated and timeless, without clocks or windows. Loss is silent, but winning is noisy: the rattle of coins from the slots, the whoops from the craps table. Food, drink, and entertainment appear as free and plentiful as in a medieval peasant's idea of heaven. In some Las Vegas casinos, there are even oxygen bars where you can recharge your tired brain. If you are someone whose pain and pleasure responses are naturally slightly muffled, all this will help bring you up to par and keep you at the tables. And if you're risk-averse, you will enjoy the perks all the more—but don't gamble: your insula wouldn't approve.

HYPERBOLIC DISCOUNTING, OR THE CURSE OF CREDIT

"I would gladly pay you Tuesday for a hamburger today"—the words of J. Wellington Wimpy sum up yet another quirk of behavioral economics. Rational thinking assumes the concept of *discounting*: the further in the future I receive a good, the less I need pay for it today. This is the flip side of interest and is usually calculated the same way, using a linear scale representing some fixed discount rate. So if I am willing to pay one dollar for a hamburger today, I should pay, say, ninety cents now to get hamburger tomorrow and eighty-one cents for a hamburger the day after.

Emotionally, though, "sometime soon" and "right now" have totally different values. Confronted by the prospect of an immediate good—the warm, fluffy bun and juicy beef; the hint, among other seductive aromas, of caramelized onion; the prospect of crisp, cool lettuce with the sharpness of well-made relish—we act like Wimpy and throw all calculation out the window. We want it *now*.

The result of this overwhelming desire is called "hyperbolic discounting": we can only judge accurately between two goods as long as both are right in front of us, or both are far in the future.[44] If one alone is imminent, it sweeps the board. Although we agree that $1.10 in thirty-one days is clearly better than $1 in thirty, we'll always take $1 today in preference to $1.10 tomorrow.[45] We even use entirely separate parts of the brain to value immediate and delayed rewards—the first through the emotionally intense limbic system and the second through the more coolly judgmental prefrontal cortex.[46] This could be sound instinct (a bird in the hand, and so forth) if it weren't for the crucial fact that any economic choice is also an implied purchase: that is, if you consider $1 today to be *worth* $1.10 tomorrow, someone is bound to try to *sell* it to you on those terms—by offering you a credit card.

You'll remember that we judge whether a price is fair by balancing anticipated pleasure against the pain of expected loss; both of these are emotions of the moment. But if the pleasure is immediate and the pain is in the future, how can we possibly compare them? This is the insidious horror of credit: by punting payment off into the middle distance, it brings hyperbolic discounting into play whenever we feel the urge to acquire. Barter is clearly better: if I want your Steve Dalkowski baseball card, I know I'll have to hand over two of my Nolan Ryans. Paying cash, too, offers a physical sensation of loss, a worrying lightness of pocket, by which to judge how rational a purchase really is. But consumer credit is the love child of corporate greed and individual irresponsibility: it removes the pain (and thus the brain) from the decision. We order the hamburger—because, basically, we think Tuesday will never come.

Christmas is always coming, though, which is why people with little cash to spare often join Christmas clubs, despite the fact that most such clubs pay no interest and many even charge a membership fee. Since we naturally find it hard to restrain ourselves, we set up these artificial, irrational deals with the self to force us to do what we will realize in retrospect we should have been doing all along. Life is full of these contracts with the self, from putting the alarm clock across the room to buying an expensive gym membership ("I *have* to go or I'll have wasted all that money"). Leonid Brezhnev, leader of the Soviet Union, controlled the military power of millions, but he knew he couldn't control his urge to smoke, so he had a cigarette case fitted with a time lock.[47] Not drinking on weekdays, finishing our carrots before we can have dessert—the mind seems a double act, with Id as the stooge and Superego as the smart operator. A Yale economist, Dean Karlan, has even set up a Web-based company to exploit this instinct by inviting people to make binding financial pledges in support of desired goals like losing weight or finishing a thesis.[48] If they don't achieve what they have committed to do, the money goes to charity—or, even more motivating, to a person they hate.

Such a scheme will not work for everyone, however, because the emotional intensity of reward and loss varies widely between individuals. For some people, a picture of delicious chocolate cake doesn't just initiate pleasant "*mmmm . . . cake*" thoughts; it activates a powerful motivational process in the brain ("*that's what you want; get it now*").[49] These, unsurprisingly, tend to be people who develop eating disorders; a similar process seems to govern drug addiction.[50] Such simple variations in the response to pleasure can make a great difference in how we find life treating us. The psychologist George Ainslie, who first identified hyperbolic discounting, said that the more prone we are to it, the less feeling we have that we control our futures. When you think how much of modern life depends on delayed gratification—from getting an education to finishing a chapter—you'll agree how important that feeling of control is. Belief that all is kismet, like reaching for the cake, can be a dangerous move.

FOR SELECTED SUCKERS ONLY

"A lie is an allurement, a fabrication, that can be embellished into a fantasy. It can be clothed in the raiments of a mystic conception. Truth is cold, sober, fact, not so comfortable to absorb."[51] Whether appearing as himself, as Dr. Weed (representative of the famed Verde-Apex copper interests), or as Dr. Reuel (associate in Kimball's confidential money-duplicating machine), Joseph "Yellow Kid" Weil was always perfectly turned out: a well-trimmed beard, glittering pince-nez, a sober suit of dark but lustrous cloth, spats, and a faint whiff of violets. He chose as his targets men who were both stingy and self-willed (and would not, therefore, consult their wives). He rarely proposed his business straight out; it needed to be dragged from him—but once it was, it always proved fascinating: battles with J. P. Morgan, secret clubs of European royal investors, abstruse points of mine-property law that would make the "shrewd" listener a great deal of money at someone else's expense. For half a century, Weil was the world's most consistently successful confidence man. He was only incidentally a liar; his real secret was to tap into deep reserves of irrationality that made his victims convince *themselves.*

Con games depend on two mental quirks for their persuasive power: what Kahneman and Tversky named the *availability heuristic* and the *representativeness heuristic.* The first makes us attach undue weight to the memorable and literarily satisfying. We prefer the explanation that makes itself most easily *available*: one good story beats a hatful of statistics. "If I can remember it, it must be important." The second makes us equate things that appear the same, regardless of their basic likelihoods. A suave, continental business manner seems just as *representative* of creditworthiness in, say, Dubuque as on the Paris Bourse. If, out of all offices in the world, an expensively dressed man walks into mine with a tale of untold wealth that only someone of my excellent repute can release, I'd need an inhuman faith in the laws of probability to stop and ask, "Why me? Just how likely *is* this?" Indeed, the more status I had and the more control I believed I exercised over life, the more natural it would seem to be offered such an opportunity—which is why boiler-room stock pushers so often target surgeons and e-mail scamsters go for clergymen.

Homo economicus, the hero of all textbook problems, is an ideal agent, showing unbounded rationality, will power, and self-interest. He considers all options, plans for the long term, rejects disadvantageous offers, and flosses thrice daily. He is, of course, a mythical being. Most of us don't know what we know or don't know—indeed, we often don't even know what we want. We are prone to fatigue and inattention; we are swayed by emotion and conflicting gut feelings—and can resist everything except temptation. How, then, can there be a science of economics? Simple, says economic theory: mistakes are made individually, but laws are made collectively, by the *market*. Multiple trading can replace our personal faults with a general intelligence, vigilance, and value-assessment. Individual investors may—will—make stupid choices, but this opens the door for *arbitrageurs*, the quick-witted agents who spot a discrepancy in price and, by exploiting it, remove it. Perfect markets should supply the world with the right value for every asset, based on all available information. In this fable, it is the crowd, not the individual, who will notice when the emperor is naked—or, as with the Yellow Kid, too well dressed.

Unfortunately, collective investment provides no insurance against the irrational effects of the availability and representativeness heuristics; indeed, it can amplify them. A Ponzi fraud is the most basic form of mutual fund, its simplicity untarnished by the need for any real assets: early investors are paid out of the deposits of later ones, while later ones will see their paper gains add up, just as long as they don't attempt to cash out. The availability heuristic naturally demands that the promoter claim some brand of novel financial wizardry; the original Charles Ponzi said he was exploiting a price differential in international postal reply coupons. More recent schemes have based their astronomical returns on promissory notes, letters of credit, matching gifts, accounts receivable, "prime bank notes," "SwissCash mutual funds," and paid Internet autosurfing clicks. Representativeness exploits the fact that all shared investment is a shared experience: "everyone" is talking about it—there are newspaper columns, cable TV channels, and Internet chat rooms dedicated to it—so there *must* be something to it. Representativeness was also the key to the many "pump and dump" schemes that punctuated the dot-com boom at the turn of the millennium: multiple, differently signed recommendations on message boards would pique general buying interest in a penny stock already owned by the recommender—who would promptly sell out, leaving the market with the poke but no pig. The most successful of these operations was run out of an upstairs bedroom in Cedar Grove, New Jersey, by a tycoon named Jonathan Lebed.[52] He was fifteen at the time; the Internet means that adulthood is now no more necessary to a good scam than spats and the scent of violets.

Yes, you say—but all this is about frauds and bubbles; real markets don't work that way. Up to a point, Lord Copper. If investment markets exist to iron out irrationality, then why have Shell and Royal Dutch Shell—twin stocks in the same company, separately traded through historical accident alone—often diverged in value from each other by up to 35 percent at any one time? Why do shares in closed-end funds, which offer investors a portion of known assets, trade at prices 10 percent above or below the value of those assets? Why does a stock gain an average of 3.5 percent simply by joining the S&P 500? Why do shares in small companies tend to do well in January and on Fridays, but badly on Mondays?[53] And why, if every seller theoretically thinks this is a good price at which to unload something, are there so many buyers? On the day this was written, the volume of trades on the New York Stock Exchange was 1,345,349,480, so 2,690,698,960 parties each must have felt they had found a bigger chump.[54]

A fool and his money are soon parted, which means that if enough of them enter the market at once, it becomes not an efficient mechanism for attributing true value but a device for impoverishing fools. This, perversely, puts the fools in charge: their habits will determine the behavior of the professionals. When something like a tulip mania arises, the wise investor does *not* smooth out this obvious mispricing—he buys tulips (at least for a while). Folly becomes wisdom; wisdom, folly: Tony Dye, the fund manager who warned us against the dot-com bubble so underperformed the market that he lost his job, mere days before he was proved right. (He had previously seen through the Japanese bubble—and would subsequently see through the housing bubble—but always that little bit too soon.)

Logic dictates that markets should be quiet places, where calm people sensibly buy into the future profits of well-managed concerns, shifting and diversifying their holdings as they monitor gradual changes in the world's patterns of consumption. Well, not on this planet, they aren't; they aren't even *random*, or we could use the algorithms of mathematical probability to understand them. The real free agents in today's markets, the hedge funds—empowered to buy long and sell short using the most frightening derivatives; all unregulated, unscrutinized, and using other people's money—still fail to generate consistent absolute value: even between 1990 and 2000, as the bull market roared ahead, more than 90 percent of them added *nothing* to their underlying market risk-return profile; and that's before charging their enormous fees.[55]

In the absence of logic, we are left contemplating a noisy, rumor-driven world, where the winners over the last three years will be losers over the next, and vice versa; where the emotional nuances of the *Wall Street Journal*'s daily "Abreast of the Market" column better determine the next day's trading than most computer predictions; and where the best time to sell a stock is when the company has appeared in glowing terms on the cover of a business magazine (yes, it works just as well—and for the same reasons—as the famous *Sports Illustrated* jinx).[56] And if you think financial markets are irrational, you haven't seen the art market, where the price of finely chased eighteenth-century silverware still closely tracks the melt-weight value of the metal, but a can (nicely labeled) containing an Italian artist's turd is worth sixty-seven thousand dollars, having outperformed gold by seventy times over thirty years.[57]

MACROECONOMICS: PUMP AND TRICKLE

Hidden away in the Slightly Dull section of the Science Museum in London, far from the model locomotives and depressingly close to the geometric solids, stands the Moniac. About the size of a fridge-freezer, it is a perplexing assembly of hoses, plastic tanks, pulleys, and the windshield wiper motors from a Lancaster bomber. It is also a machine for running the U.S. money supply.

It was built in 1950 by A. W. Phillips, a New Zealand–born electrical engineer and tinkerer (he had built both a miniature radio and a clandestine electric teakettle for his POW camp in Java), known to economists as the deviser of the Phillips curve that relates unemployment to inflation. His machine, essentially a nonlinear analog computer, tracked the movements of an alarmingly bloodlike liquid currency through the economy, from a GDP pump at the top to a national expenditure sink at the bottom. Mutually adjusting servos and valves in the middle dribbled money around various tanks: Federal Reserve, taxation, savings, investment, foreign exchange. Pens attached to floats generated time graphs of each important variable. It's both impressive and oddly reassuring: all you need to do is dial in your rates, measure out a jug of currency, switch the thing on—and all the complex behavior of a real economy magically appears, bubbling and gurgling.

The Moniac became a highly successful teaching machine; there are still examples at Cambridge, Harvard, Ford's world headquarters, and the Central Bank of Guatemala. It reassured students that the torturous formulas they had to learn actually did describe something in the real world. Less usefully, it created a sense that all this could go on without human agency: that people's hopes, beliefs, prejudices, and fears were, once aggregated, just a subdiscipline of hydraulics.

Science ought to be unsentimental, so perhaps this assumption is right—except that it generates some *further* assumptions that simply don't work in real life: not just local matters of irrational behavior but also big, macroeconomic anomalies. George Akerlof (like almost every other economist mentioned in this chapter, winner of a Nobel Prize) described five of them in his presidential address to the American Economic Association in 2007.[58]

First: classical, utility-maximizing economics would expect people to increase their consumption only when their wealth (that is, the total of current income plus discounted future income) increases. But they don't; around half of us blithely increase our consumption when only our *current* income has increased, without even thinking about the future. For every thrifty ant in society, there is always at least one feckless grasshopper.

Second: if you and your children were just tanks in the Moniac, you would do best to empty yourself of any state handouts (through, say, riotous living) before you die, because whatever the government has given you through social security will eventually have to go back by way of inheritance and capital taxes on your children. Why pass on this essentially borrowed money as a bequest? Yet people do, scrimping through their last years just for the pleasure of having something to leave.

Third: company managers should, in theory, maximize value for shareholders through well-judged investment. Whether they do this using available cash flow or through buying back shares and issuing debt is immaterial; if it's the right investment, the funding options should balance perfectly. Observation, though, shows that, as with individuals, current income always beats long-term calculation in prompting investment. When the company is flush with cash, it seems managers just *have* to buy something.

Fourth: the Phillips curve relates inflation and unemployment pretty simply: the lower unemployment is, the higher the rate of wage increases. The idea has been tweaked since to include a "natural rate" of unemployment, below which increased wage expectations set off an inflationary cycle and above which deflationary expectations send us tottering off toward recession. But this can't explain America in the 1990s (low inflation at historically tiny rates of unemployment), nor the fact that, in depressions, wages tend not to drift down—it's just that fewer people get them.

Fifth and last: economics is based on rational expectation;

expectation should include, for instance, what you think the Fed will do next month. So if there's high unemployment and the bank decides to address this by easing monetary policy, price and wage-setters across the economy should anticipate this by raising their demands. All that will result is inflation. In theory, therefore, monetary policy will be ineffective in stabilizing the real economy; but observation shows times of markedly greater or lesser central-bank power. In the 1970s, Fed chairmen Arthur Burns and G. William Miller grappled violently with stagflation, to no effect; in the 1990s, Alan Greenspan could settle the markets with a lift of his eyebrow.

Akerlof's point is that, in each of these five areas, classical economics has left out the concept of the *norm*: the social constraints that keep us from acting any old how. Even in aggregate, people have assumptions about the right way to behave—assumptions not purely rational. For consumers and corporate managers alike, income is fundamentally different from capital; it is *right* to spend it, whether on buying a better car or building up a departmental empire. Dipping into savings or incurring debt, however, would be wrong. It is *right* to leave money to your children, just as it is right that they should live in squalor when they're students and achieve their own financial independence as adults, no matter how large their inheritance. It is right to get a wage increase, but it still counts as an increase even if it only matches the rate of inflation. It is right to pay the same for steak this week as you did last, regardless of what happened to oil prices. And it is right not to spend all your time thinking about money.

These norms combine to prevent real life from imitating the model. The servos of the Moniac adjust flow smoothly and constantly in response to incremental pressures, but a system with people in it is bound to have sticking points. We are parents and children, workers and managers, taxpayers and consumers: every role has its own rules. We are never simply "economic agents."

ONLY THE FAIR DESERVE THE GRAPE

Here's another snapshot from the busy world of the capuchin monkey, proving that you don't even need to wear clothes to have a strong sense of the norm: at the Yerkes National Primate Research Center in Atlanta, a group of brown capuchins have taken up our daily habits of work, pay, and expenditure.[59] They are by nature a cooperative lot, sharing both tasks and food. But when it comes to rewards, they have a keen eye for the deal. Grapes are good; cucumber slices less so. Give a female capuchin a cucumber slice for her hard-earned token, having also given her neighbor one, and she will settle down to eat it, albeit not with much zest. Do the same after having given her neighbor a *grape* and she is likely to fling the cucumber at you.

It's never just about the cucumber, is it? Kaplan's Law (Ellen, as a genuine grandmother, has the right to make such laws) says that of the three things we work for—money, the interest of the job, and the respect of our peers—any *two* will do. Just one is not enough. No salary will be enough to compensate for a life of contemptible busywork; accountants may be the butt of popular jokes, but they are paid well and the work appeals to certain neat, analytical minds. Teaching is interesting, but as teachers lose the automatic respect of society, they tend to want more pay. Even pigeons, not usually considered paragons of economic sophistication, have worked out that some jobs are not worth having: faced with having to peck a bar hundreds of times for a meager reward, they will instead peck a bar that renders the first one inoperative, in effect firing themselves in a style reminiscent of country and western songs.[60]

We have something else in common with the birds: we flock together. The abstract theory of compensation assumes that we exchange personal labor for personal pay, but our actual norms are different: even the loneliest freelance Web designer, once employed, becomes temporarily part of a social enterprise, where notions like *fairness* and *purpose* can skew the calculations of utility.

At the University of Zurich, these very notions are the research topic of Professor Ernst Fehr (it would be nice if his colleagues were Professors Weise and Gut, but in fact they are Fischbacher and Kosfeld). The Fehr laboratory specializes in studying social choices, using two economic conundrums originally devised at the RAND Corporation during the cold war: the prisoner's dilemma and the ultimatum game.

You may be familiar with these, but if not, here is a brief summary. In the prisoner's dilemma, you and another (usually someone unknown to you) are assumed to be captured by some authority and each in a position to betray the other. If you squeal on him and he stays silent, you go free and he goes down for a long stretch; the opposite but equivalent applies if he rats on you when you stand firm. If you betray each other, you will both serve a medium sentence; if you can cooperate and stay silent, you both serve a short sentence. Game theory concludes that the best strategy for an individual (that is, the strategy that secures the minimum jail time no matter what choice the other makes) is to betray the other, which might explain why the cold war lasted so long.

The ultimatum game offers you a sum of money to divide between you and another player—but only if you can get the other player to agree to the division. Game theory claims that he should submit to any division, since something is always better than nothing. You should therefore offer the minimum, since this means you will get more. Computers (and, it seems, people with autistic spectrum disorders) find it easy to behave like this.

Professor Fehr's subjects, though, do not.[61] In game after game, we see them cooperating with unseen fellow prisoners and insisting on equitable division of ultimatum money. When given the chance, they go even further, willingly paying part of their own goods to punish other players whom they see as taking unfair advantage. Nor is this a matter of securing long-term stability or preserving a reputation for honesty: the same good-citizen qualities appear when the game has only one round and when the players are mutually anonymous. For about half of the subjects, social norms consistently rate as more important than individual gain.

Before you say, "Well, yeah—but this is a bunch of Swiss graduate students," Professor Fehr has anticipated your objection. He ran a similar game in Moscow with *real* money—three months' median local salary—at stake. He has also taken his dilemmas to Austria, Germany, Hungary, the Netherlands, and the United States. In each case, the same proportions of social cohesiveness to selfishness appear. Other researchers have extended the study to groups in places untouched by economic or psychological theory: to the Hadza of Tanzania, the Torguud of Mongolia, the Gnau of Papua, Aché of Paraguay, and Machiguenga of Peru.[62] Whether foragers, farmers, shepherds or gardeners, villagers or nomads—Turkic, Cushitic, or Macro-Panoan Isolate—these people showed a uniform preference for the social norms of fairness over individual self-interest. Wherever *Homo economicus* lives, it is not on one of the inhabited continents.

All people are social, but admittedly some are more social than others: play the ultimatum game among the tightfisted Peruvian Quichua and you will come out with far less than among the openhanded Lamalera of Indonesia, who seem to think it shame merely to offer half. But then the Quichua are willing to accept whatever *you* offer, while the lordly Gnau will turn down positive offers more frequently than Pittsburgh college students. The variation itself seems significant: generally, the more mutual dealing a society has, with more integrated exchanges of benefits, the higher the importance of social norms. Individual variation is entirely insignificant; the standards of the group prevail. Perhaps most important, none of the groups thought these sharing games were some mad amusement of professorial outsiders: they recognized that deciding the proper distribution of goods is the stuff of everyday life. (The Orma of Kenya rather offhandedly mentioned that they had the same game themselves, but called it *harambee*).

What is behind all this? Is each of us constantly, but unconsciously, calculating long-term benefit against local advantage? Are we keeping a running total of favors traded, so we can parlay our good reputation into future wealth? Are we perhaps securing advantages for our offspring—or is some unselfish Helvetic gene using us as its vehicle toward world domination?

Apparently not. The careful design of the studies has made it possible to discard most such "purposeful" theories, and what remains is something Fehr calls *strong reciprocity*: an intrinsic human willingness to sacrifice personal advantage in order to reward the kind and punish the unkind. It is, and probably always has been, the force that makes informal economic life possible. It allows us to make surprisingly trusting contracts with complete strangers: hop in a taxi in Lagos, Jakarta, or São Paolo and the driver will take you where you want to go before knowing whether you have the money—and you, most likely, will pay rather than skip on the fare, even though you know he's unlikely to leave his cab to chase you. Strong reciprocity also governs employment: Fehr has found people universally offer to work harder than the minimum required (except, of course, at late-night video rental stores) while employers reward extra effort at more generous terms than the base rate of pay. The rhetoric of class warfare, logical though it may be, runs aground on this irrational class pacifism: boss or labor, when we work at the same place, we believe we are working together.

Similarly, our hatred of being cheated is such that we can make the ultimate sacrifice for it: think how many old ladies are killed clinging to purses with less than fifty dollars in them, or people who are run over protecting cars on which they have paid theft insurance. Pure, dishonest selfishness is a shock because we expect, on the whole, fair dealing. Yes, under sufficient pressure, the bonds of fairness can dissolve and man becomes a wolf to man—but this happens surprisingly rarely. We are all more than a little Swiss.

Is there an even simpler, deeper urge beneath this strong reciprocity? Jeffrey Goldberg, Lívia Markóczy, and Lawrence Zahn of the

University of California think so.[63] In their studies of the prisoner's dilemma, they isolated two qualities that could explain why we tend to cooperate with others when logic says we should betray them: symmetry and the illusion of control. That is, we believe, without evidence, that others will think as we do in certain situations, and we overestimate our influence on events where we have a choice. Consider voting: are you shocked and annoyed if you find the majority is not of your opinion? Do you vote because "if I don't, who will?" Yet logic tells you that plenty of dolts make it to the polls and that your vote among the millions means nothing. Similarly, while game theory makes clear that being a selfish traitor is the logical choice, symmetry makes us feel that others, like us, would wish to do the right thing; simultaneously, the illusion of control tells us that if we are good, our goodness will somehow influence the situation to come out right.

Habits and choices may, of course, be a matter of social training—when we see equivalent behavior in a Mongolian yurt, Amazonian leaf-hut, and fluorescent-lit seminar room, it may simply mean that the pressures of social life (like the proverbs of mothers) are remarkably similar around the world. And yet there are some physical experiments to suggest that the assumptions behind our social norms are more hardwired. Brain imaging has shown, for instance, that the dorsolateral prefrontal cortex, part of the frontal lobes, becomes much more active when a subject is faced with an unfair offer. When Fehr's colleagues used trans-cranial magnetic stimulation (strong magnetic pulses produced by coils held over the scalp) to dampen activity temporarily in this area, then presto! Subjects willingly accepted deals they had found insultingly low before.[64] Other studies have shown that punishing cheaters activates the same areas of satisfaction in the striatum as does receiving a cash payment, so while we lose economically if we are paying to punish, we gain exactly the same satisfaction as if we had made a profit.[65]

Folk wisdom would assume that men and women do not respond in the same ways to such powerful matters as cooperation, betrayal, and revenge—and in this case, folk wisdom is right. Testosterone levels are a good predictor of how likely a subject is to refuse an unfair offer ("*Three* lousy bucks? Take *that*!"). But men's brains tend to switch off after an economic decision is made, while women's show continued activity: anticipating future reward, planning strategy, resenting unfairness.[66] Women's brains, when they cooperate, light up in the reward region of the striatum and the learning-by-reinforcement centers of the orbitofrontal cortex, suggesting that they think doing the right thing is not only more satisfying than selfishness but also fits better with experience—it's more *natural*.[67] This last result finds some confirmation in the real world: another Nobel laureate, Muhammad Yunus, gained his prize for setting up the Grameen Bank in Bangladesh, a microcredit institution that supported development projects among the poorest people. Its technique is to advance money to groups of women who accept collective responsibility for the loans to each—if one is in default, the others cannot borrow more. So why does Grameen choose primarily women? "They are simply better with money," says Yunus; by which he means they are better with credit—which, after all, literally means "trust."[68]

This discrepancy between the sexes also suggests an interesting conclusion from one of Fehr's results: when around 40 percent of a group is made up of strongly reciprocal people, the norms for the *whole* group become strongly reciprocal. Women, as we've seen, find cooperation more pleasurable than men do; they tend to favor fairness rather than apportion bragging rights by putting one over on the other guy. So the more women there are in a group, the higher the chance of reaching that critical 40 percent. This implies that, since reciprocity is the key not just to good manners but also to the smooth and efficient working of an economy—where we can all get to the job at hand without the constant irritation of cheating and being cheated—then the speedy promotion of women in developing countries is not merely a moral good and a spur to the gentler arts but an economic imperative as well. That, at least, is the authors' opinion—but then, like everyone, we assume that everyone thinks like us.

SELF-SERVICE

One of the oddest places of work ever devised must be the *sharashka*: an institute/prison within the Soviet Gulag system. Here engineers, designers, and other experts would labor, unpaid, in complete isolation, serving the people of whom they were officially enemies and the state of which they were officially wreckers. Aleksandr Solzhenitsyn worked in one, which he describes in his novel *The First Circle*. The aircraft designer A. N. Tupolev created the Soviet Union's World War II bomber fleet from a *sharashka*; when he needed more draftsmen, the secret police obligingly went out and arrested them for him.

While Russian culture treasures irony as much as it does self-sacrifice, this hideous system seems to push both far beyond the range of credibility. Nevertheless, Stalin had nothing to fear from his prisoner-professionals: the desire to accomplish something through collective effort outweighs any qualms about whom the work will ultimately benefit—or else there would be no nuclear weapons designers, cold-call marketeers, or tax collectors.

The culprit here is the *self-serving bias*, a powerful identification with Our Side, no matter what that might be. Its effects can be startling: when Princeton and Dartmouth students were asked to review film of a Princeton-Dartmouth football game, the Princeton students saw three times more fouls committed by Dartmouth than the Dartmouth students did.[69] When law students were asked to evaluate and attempt to gain a settlement in a mock tort case, those who read the case before being assigned to one party were able to reach a settlement 94 percent of the time; those who already knew which side they represented could only manage it 75 percent of the time—despite the fact that they would lose out if they could not settle.[70] Self-serving makes us identify with our employers in odd ways. If I work for a power company, complaining customers are Them, the selfish, lazy, late-paying people who make my work difficult. When my car breaks down, though, the insurance company becomes Them, the greedy and complacent fat cats who dispute my legitimate claim.

Companies can be more or less successful in their understanding and manipulation of the self-serving bias. L. L. Bean offers to guarantee its goods unconditionally, potentially opening itself up to an army of people demanding replacements for worn-out duck-hunting boots. But in doing so, it artfully brings its customers onto its side: "We are the kind of people who choose to buy from such a trustworthy supplier; we, therefore, will also be trustworthy and not abuse the guarantee." Contrariwise, a company such as Wal-Mart, which for years acted as a purely economic unit, rigorously slashing costs and prices with no more implicit promise than to give people what they want for less, almost invites the enmity of all—because it has, until recently, failed to *pretend* to be more than a business.

BUT FIRST, A MESSAGE FROM . . .

Rational economics is about choice, and choice is based on knowledge. The utility we seek may be subjective (what's good for me), but we are assumed to know ourselves well enough to gauge our wants correctly. Now then, what would you say to a glass of Château Latour '61? "Yes, please"? Well, how do you *know* you want it? Not from experience, unless you are extremely fortunate. No, the knowledge on which you based your judgment is *hearsay*—which explains why advertising is such a huge business.

Sadly, most things advertised lack the obvious, heady, ruby-tinted desirability of Latour. Merely letting the public know of your willingness to sell compact cars, budget hotel rooms, or cheap pants is unlikely to activate the "*yum*" response that, as we have learned, underlies most purchasing decisions. People need to be inveigled into linking the mundane with the irresistible; clever advertising, therefore, depends on association to make the things we may merely need appear also to be what we want—thus disguising something of economic value as something of personal worth.

During the frothy finale of the big bull market, the *Financial Times* put out a magazine called *How to Spend It*, whose purpose was to suggest directions in which hedge fund managers and their spouses can shovel out enough cash to keep their heads from hitting the ceiling. A fabled resort in Mauritius found it the ideal medium for a double-page ad: "A glass of lychee rum on cracked ice; ghost crabs on the sand after sunset; frangipani flowers on the surface of the pool; reflections dancing on the side of a speedboat." The real message, though, was the image: from mid-thigh to navel, the tanned, glistening loins of a woman apparently intent on shimmying out of her thong. *Come here and get this.*

You will not be surprised to hear that our simian cousins go in heavily for this genre of advertising. In a study that let monkeys pay (in sips of juice foregone) for the chance to look at various pictures, male rhesus macaques were willing to give up any amount of juice to keep ogling snaps of female macaque bottoms.[71] The surprise is that they were similarly entranced by the faces of high-status males, while they had to be *paid* to look at low-status group members. We cannot know whether the macaques felt the same about alpha males as we feel about, say, the Marlboro man or the athletes who endorse Nike, but it is clear that status and lust, sex and the gang, can fundamentally disrupt any primate's sense of personal advantage.

The economist Marianne Bertrand and her colleagues have measured just how much this disruption is worth. They set up an experiment in South Africa in which a local lender sent out fifty thousand letters offering large short-term loans at randomly chosen rates of interest ranging from 3.25 percent to 11.75 percent per month.[72] The lender also randomly assigned a variety of "marketing treatments," ranging from detailed descriptions of the loan terms to promotional giveaways. The relative take-up of the offer at the range of given interest rates would then show how successful each marketing technique was. Among the most powerful persuaders for male loan applicants was simply a picture of an attractive woman in the corner of the letter; it produced an increase in take-up equivalent to dropping the monthly interest rate by 4.5 percent—worth, on the average 1,000-rand loan, an extra 540 rand per year to the lender. Supermodels do indeed earn their huge fees.

This blinded-by-sex effect is reproducible in the lab. Research at the University of Leuven in Belgium found a significant correlation between the ratio of finger length (index finger to ring finger) and men's willingness to accept unfair offers after being shown pictures of sexy women or lingerie.[73] Hmm? *Finger* ratios? Why, yes: the shorter your index finger is in proportion to your ring finger, the higher your exposure to testosterone before you were born. Higher testosterone, as we have seen, generally makes men less willing to be imposed upon in an ultimatum game, but apparently it also makes them more distractable: sex cues make them more likely to accept "unjust" divisions and more interested in immediate, smaller rewards; the prospect of passion cranks up the hyperbolic discounting. Thus London merchants long ago learned to hire pretty shopgirls and auto tool companies discovered the power of the cheesecake calendar: sex sells, at least for men, because it knocks out the capacity to spot an unfair deal.

Advertising, as its name suggests, began with apparently useful information about forthcoming books or patent medicines ("ask your apothecary if henbane is right for you"); and, just as men are not the only consumers, sex and status have not been the only messages. Trustworthiness is also important; in a world with many dangers, we respond to the cues that signal a safe, dependable experience—even preliterate toddlers much prefer to eat McDonald's food from its branded container than from an unlabeled package.[74] We may think that we choose a particular brand consciously as a rational agent, weighing its advantages, but the truth is usually different: a familiar logo saves us from the effort of decision. Companies like Procter and Gamble built their success on the fact that no one wants to *think* about what toothpaste or laundry detergent to buy.

We say, in polls, that we want information about products, but this is only partially true. The more information we get, the harder we find it to commit to any product, because we feel that if we could get even *more* information, we might find something even better. The computer industry has both thrived and suffered as a result of this compulsion, as products quickly win and lose market dominance based on essentially arbitrary metrics: "64x read/write drive! Blazing 1.2 gHz clock speed!" How much are these desirable but mysterious advantages actually worth? For obsessive video-game players, perhaps something like their market price—but for the rest of us, the One Laptop Per Child project has demonstrated that a machine doing almost everything we need can cost just $173. So at least two thirds of what you spent on your last computer went just for the sizzle.

We say we want to be warned about future dangers, from the hazards of secondhand smoke to the halitosis even our best friends won't mention. In fact, we don't. Such commonsense matters as AIDS or cancer testing show the same odd discrepancy: people want the testing to be available but do not go to be tested themselves. Similarly, public service messages have to be much more powerful than commercials to have an equivalent effect—or need to substitute some positive good for a negative message. Looking at children, as we know from the baboons, is a positive good—and that's the reason you never see poor *adults* in those late-night TV messages from charities. We primates respond badly to images of low status.

Choice, then: the obvious point of advertising must be to let us in on the myriad options we have; to remind us of how lucky we are not to live under the state-directed drabness of communism, where all groceries came from the House of Food, all headwear from the House of Hats. Well, choice turns out to be more honored in the breach than in the observance. Even doctors can find it paralyzing: when faced with a choice of two treatments to prescribe for arthritis, a significant proportion failed to prescribe either. Bertrand's South African lending experiment found the same; giving a choice in the offer letter had a *negative* effect equivalent to 2.3 percentage points in interest. This may be why many cell phone operators offer a labyrinthine series of tariff plans, knowing full well that people will stay in the most expensive bracket rather than have to face the agony of choosing. Our ideal of practical life remains Kodak's promise: "You push the button and we'll do the rest."

If it's not information, warning, or choice, what then are we looking for in the thousands of hours we spend in front of advertising? What is the point of the half-trillion dollars the industry spends globally each year? Three simple words: *recognition, prestige*, and *association*. Let's start with *recognition*. The mind aims for efficiency, which is one reason we dislike choice; if we can come up with a single recognizable brand when we consider a need, we will favor that brand. In brain-scanning experiments, people much prefer the taste of Coca-Cola *when they have seen the brand* to when they have simply sipped it from an unlabeled cup (curiously, Pepsi does not show the same effect).[75] This one-choice bias probably explains why, despite the efforts of armies of corporate lawyers to stop us, we still say "Xerox" and "Kleenex" when we mean their generic equivalents.

Prestige, to a degree, is obvious—if Andie MacDowell really thinks you're worth her brand of makeup, it would be churlish to disagree. Endorsements offer entry into the top circle, the promise that we, too, can become the kind of people our fellow primates would pay to look at. And it's not only individual status that advertising implies is transferable—class distinctions, subtle but significant, pervade the medium. Ralph Lauren made his fortune selling upmarket versions of the chino pants and chambray workshirts that preppy men used to buy at the hardware store; his ad campaigns still breathe a sense of sun-burnished, salt-tinged afternoons on Vineyard Sound, his models gawky and rawboned but utterly self-assured—as only Groton and

Miss Porter's can make you. Even the slightly ill-fitting quality of the clothes is part of the message: a meticulously constructed image of people far too confident to care about appearances.

In the absence of class, *price* is always a good guide for the uninformed—again, because we conflate value and worth. As the Russian joke has it, two *biznesmeni* meet: "Yuri! Nice tie!"; "Versace—it cost me six thousand rubles." "Are you crazy? I know a place you could get that for *twice* as much!" Paying a lot for something not only shows that we *can*; it displays our willingness to accept the standards of the group—we choose based on a shared opinion, expressed in price. Thus people at a blind wine tasting, when told that a wine was expensive, routinely rated it higher than when told it was cheap. But when *un*aware of the price, they actually tended toward the $2 plonk rather than the $150 tipple.[76] Thus a lot of prestige advertising will emphasize exclusivity—the unwillingness of the vendor to sell to just anyone. Only those with exquisite taste can be allowed to spend (as one lucky buyer recently did) fifteen million dollars for a tank of formaldehyde with a dead shark in it.[77]

Prestige, like price, is a relative matter. H. L. Mencken defined a wealthy man as someone who made more than his wife's sister's husband.* Nothing is worse than being subject to an unfavorable comparison: stroll down the shopping streets of the Hamptons, St. Tropez, or Key West and you will see people caught up in the same anxious routines of competitive display and consumption that they had originally come to these once-simple fishing villages to escape. Do you have the season's must-have? In this context, it really matters

---

* Mencken was onto something. A study from the University of Bonn reveals that the greatest activation of the brain's reward centers depends crucially on a *comparison* with the rewards received by others—and that the pain of having less is greater than the pleasure of having more. This may explain why, despite fifty years of sharply rising incomes in the West, people report themselves no happier. (See K. Fliessbach et al., "Social Comparison Affects Reward-Related Brain Activity in the Human Ventral Striatum," *Science*, vol. 318, November 23, 2007, pp. 1305–8.)

whether your black T-shirt is J. C. Penney or Armani: brand is a reassurance that, in life's division of spoils, you got the grape and not the cucumber.

Economic value reflects supply as well as demand, so the *rarity* of something can also determine its prestige and desirability. Snow was one of the most sought-after products in ancient Rome's sweaty summers: anyone with a refrigerator now lives in a way the Caesars might envy—but somehow it doesn't *feel* as luxurious. This sense that something not easily available must be worth having is the basis for all limited editions and once-only offers. It's a robust effect and one that, in a world of increasing demand and dwindling resources, bodes ill for the future. In a worrying experiment conducted at the University of Paris, shoppers at a supermarket and guests at a champagne reception alike said that they much favored a sample of "rare" caviar over one of "common," though both samples were in fact the same.[78] We notice and respond when good things play hard to get. Lobster was once used as fertilizer, mahogany for packing crates; as we consume our resources into rarity, our desire to consume more *increases*.

Advertising sells us what we know, what we desire, and what the in-crowd seems to have; but most of all, it sells us *itself*—the simple fact of repeated communication. Advertising primes our expectations; it actually shapes the world we think we see. Even if we know a message is false at one hearing and that we are being manipulated, we will accept it as true if we hear it often enough—it takes too much mental effort to resist.[79] Repetition, therefore, is one of advertising's most effective weapons. Burma-Shave's roadside jingles offered amusement for travelers in an age before the car radio, but they were also, critically, repetitive. The most irritating ads are often the most successful, for the same reason; America had not even known there was such a thing as "ring around the collar" before Wisk's hideous *nyah-nyah* jingle hit the screen in the 1970s.

Less infuriating but equally memorable advertising relies instead on conveying genuine aesthetic pleasure that, for all its slavery to

Mammon, still partakes of the qualities of art. "Use Ajax (ba *bum*), the foaming cleanser (ba ba *bum*ma bum bum)" may not have the majesty of a Mahler symphony, but it captures perfectly an optimistic jauntiness that lifts us, as we hum, briefly above this world of caked-on grime. We feel, in a limited but genuine way, our troubles float away right down the drain.

Being attractive, belonging, knowing what's what, striding through the world, whistling a happy tune: the most powerful advertising sells us the joys we already treasure, while subtly giving them a brand—twenty-eight seconds of pure emotion, followed by two of product placement. That's how *association* works. What father would not treasure the attentive affection from his son that apparently comes as standard when you acquire a certain make of Swiss watch? What mother would not want her family laughing joyously around the table, as they will if she shops for the right turkey burgers? Who would not wish to find true romance, even if he had to buy a hatchback to do so? The pleasures promised in ads are *real*, and often the most desirable pleasures we know: love, friendship, security, respect, freedom, calm, and ease. We are *right* to enjoy a good ad as a brief vision of a better, more enchanting life, all lychee rum and ghost crabs. The fake part is the belief that some product will deliver this life to us.

THE BEST THINGS IN LIFE AREN'T THINGS*

Two more studies and we are done with this vexed business of wanting and getting. Leaf van Boven and Thomas Gilovich looked into the satisfaction people feel after purchases and found a remarkable dichotomy: by and large, *stuff* does not make us significantly happier.[80] The new car, the snow blower, the elliptical trainer (*especially* the elliptical trainer) leave us, after a brief surge of enthusiasm, no more pleased with life and ourselves than we were before. Given what we've paid for

---

*We'd love to take credit for this, but it's rightly due to Art Buchwald.

them, we may be even less happy. On the other hand, the *experiences* we buy can make a permanent difference in satisfaction. A day's skydiving lessons, trekking in the Sierra Nevada, canal-boating down the Loire, even going out to the ball game—each adds to the stock of happiness in ways that do not depreciate. Why this difference? Nothing is proven, but there are four credible assumptions: an experience becomes a permanent part of your identity—*your* day on the river—while a possession may always fall away from you through loss, decay, or increasing dowdiness. Experiences are social, strengthening bonds of trust and pleasure with those we love; stuff is *yours*, which means it's not anyone else's to share. Experiences deflect envy better than objects do: I may laze on the beach at St. Barth's while you explore the cathedrals of Tuscany; neither of us really wants the other's vacation, but your Rolex is a constant affront to my Timex. Finally, experiences *slow time*; the new Mercedes merely accompanies you through the tick-tock of busy life, its successive ten-thousand-mile inspections reminding you of how quickly the years are passing. The memory of midsummer sunrise on Mount Katahdin, though, teaches a different lesson: that a single moment can contain all the exaltation of being fully, gloriously alive.

Money can buy good experiences, but it can also prevent them. Kathleen Vohs and her colleagues found that even simply *thinking* about money can stop people from behaving in cooperative ways, or from asking for help in difficult tasks.[81] A few minutes of unscrambling money-related sentences seems to put us into a self-sufficient, almost miserly frame of mind, taking us away from more transcendent values like community, affiliation, and spirituality. Perhaps the literary cliché of "poor but warmhearted" has some grounding in fact.

What it comes to, therefore, is this: gaining money, buying and trading goods, all activate the same regions of our brains as do the real and significant pleasures of life—having fun, satisfying curiosity, being sociable, doing the right thing. As a result, we too easily confuse the means with the ends and believe we can manipulate our deeper wants and fears by means of these glistering tokens. If you met some guy at a barbecue who told you that, thanks to his clever design, his house enjoyed a higher amperage of available electricity than yours, you'd be unlikely to head home and start ripping out the wiring—you'd probably edge away, nodding, toward the beer cooler. If he was talking about investments, though, you might well leave early, get on-line, and start rearranging your portfolio. Yet both topics are the same: they simply describe methods of conveying, conserving, and applying power.

Most people who become very, very rich have effectively *seen through* money early—it does not intrinsically interest them. Rockefeller was actually interested in monopoly leverage, Carnegie in vertical integration, Ford in vertical integration, Gates in . . . monopoly leverage. Each understood the dynamics of money and the role it played in realizing their wider ambitions, but once they had amassed great piles of it, their prime concern was to give it away.

Adam Smith would have approved; we should not forget that he was a moral philosopher first, an economist second. His motive in proposing freedom of trade and full play for self-interest was to clear away dogma, not to create a new one. Before Smith, it was natural and common to forbid people from pursuing a trade—and then give them a pittance in the name of charity because they were paupers; or to insist that bread can have only one price and then lynch the baker when famine emptied the granaries. Smith's target was not irrational economic theory; it was the misery brought about by conflating economics with wider moral and social issues, forgetting that goods and the Good only intersect at a few points. He would probably be appalled to see how some present-day followers have elevated his ideas into a creed, with the free market as a desirable end in itself rather than, as he saw it, simply a more efficient vehicle for the pursuit of happiness.

Of course, we do all have to make a living, so we cannot take *too* Olympian a view of lucre. Fortunately, we still have the rich to teach us how to make and keep the most of what we have. Most surveys of wealth management reveal that F. Scott Fitzgerald and Ernest Hemingway were both right: the rich *are* different from you and me—and it is *because* they have more money. Rich people save a higher proportion of their income than the rest of us and can afford to diversify their holdings more widely, insulating their fortunes against local fluctuations. Staying rich owes far less to insider stock tips murmured on the ninth green than it does to the simple laws of probability. If you want to add similarly to your smaller nest egg, the rules are simple: save more, starting earlier, than you really feel like. Don't mortgage your life for stuff—or at least buy only things that will improve with age, as you do. Beware new paradigms: "this time it's different" is true only until it isn't. Don't bet the ranch on some mysterious high-return investment that appears to be exempt from the laws of probability—the spirit of Ponzi lives on, plausible and well dressed, equally at home on Wall Street as at your country club. Put your savings *regularly* into low-cost market-tracking mutual funds, mostly equities but including some fixed-interest and overseas exposure—then forget about them and get on with life. Anxious, emotional involvement with money simply amplifies the various biases, illogicalities, and anomalies from which all those economists derived their Nobel prizes. It's only among the real pleasures—a letter from an old friend, a trip to the farm store, the first slug of iced coffee on a sweltering afternoon, sitting together watching the clouds—that we really know what we're doing.

CHAPTER III

Tinted Glasses

Thinking is what the brain does, not what we *think* the brain does. And what the brain does is massively parallel, reacting to stimuli simultaneously at the instinctive, emotional and rational levels. It throws out most sense-information and instead offers our conscious minds a summary, with helpful highlighting and foregone conclusions. Our senses are designed for being human with; our assumptions are based on the life pattern of a vulnerable but many-wiled mammal in a world full of treats and dangers. We pay attention, remember, and communicate in ways that only incidentally reflect how things really are. We are kept interested through a synthetic concept—meaning—that our unconscious brains produce for us and that we cannot help but consider the greatest treat of all.

The astonishing tangle within our heads makes us what we are.
—COLIN BLAKEMORE, *Mechanics of the Mind,*
BBC Reith Lectures, 1976

IF you fret that the world grows short of genuine wonders, consider this: the most complex lump of matter in the universe. It works in ways we can only guess at. Through generations of intense study, scientists have at last come to understand some of its local mechanism, but the connection between local and general remains for them, as for the rest of us, a matter of arm-waving speculation—we know less about what's going on inside it than we do about the functional structure of the most distant galaxies. It weighs a little over three pounds and is the consistency of toothpaste; you're carrying it between your ears.

The brain merits its superlatives because it embodies something it itself cannot handle: very large numbers. There are roughly a trillion neurons in an adult brain, each one connecting with up to ten thousand others. A thousand times more signals are made in the brain every second than there are words spoken in every international telephone call in the world in a year. Whitman was right: I contain multitudes.

Of course, this is the same brain that forgot the dry cleaning three times last week, that blanked on the principal's name at the PTA meeting, and that will never completely understand the thoughts of other brains, even those it loves the most. We may ascribe our faults in perception, flaws in character, our crimes and misdemeanors to fate, society, upbringing, or diet, but the immediate agency is always the same: for good or ill, our brains made us do it.

The two greatest problems for investigators have always been the brain's inaccessibility and its stubborn refusal to resemble any other mechanism. Snug in its bony box, it conceals the complexity of its internal connections in a bland succession of rosy folds. The impressive Latin names of cerebral features reveal in translation the bafflement of the anatomists who first discovered them: "black stuff," "grooved bit," "breast-shaped things," "uncertain zone."[1] Even where, after many years' labor, researchers have managed to map the physical connections behind a mental phenomenon, the result lacks any of the condensing power of, say, the discovery of the circulation of the blood or the electrical nature of nerve signals. Our best-documented capability, vision, follows sinuous pathways that cross and recross the brain, dipping through eight cortical regions, fanning out to the margins and doubling back on themselves like hounds on a faint trace. Looking at this spidery wiring diagram, the layman might be excused for asking, "Well, now that I know that, what do I know?"

Tangled reality demands a simplifying metaphor—and there have been many, from Plato's cave, where we sit watching on the wall the imperfect shadows of a higher world of Forms, to Descartes' theater, where we have slightly more comfortable seats but an equally distant

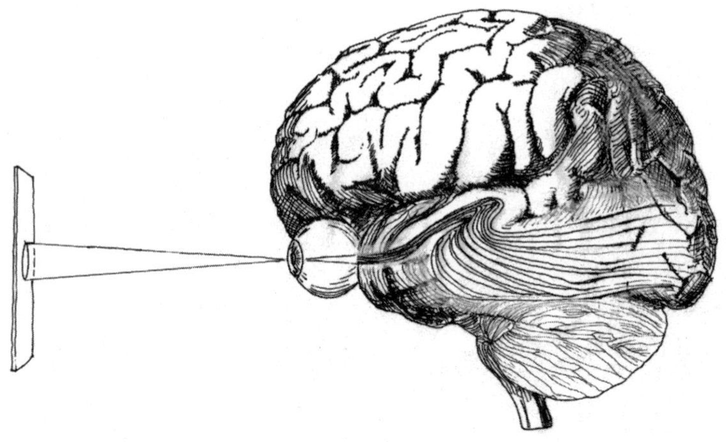

connection to whatever "out there" has written the play (the theatrical image is lent to Descartes by the modern philosopher Daniel Dennett; Descartes himself posited a spiritual observing homunculus resident in the pituitary gland). Subsequent technical fashions have generated models of the brain as a printing press, post office, or telephone exchange. The most pervasive recent metaphor has cast the brain as a general-purpose computer, with input from the senses, a memory function, and output through behavior.

Here the layman perks up: we know at least *something* about computers. Even their failures—their sudden collapses, their poor social skills—give them a recognizable semblance of character. So the brain is like a computer; well, what sort of computer is it?

A very odd one, apparently: at once spectacularly powerful and remarkably limited. Kwabena Boahen at Stanford University is attempting to reproduce the brain's neuronal structures in silicon, and he is well aware how difficult the task is. "The brain," he observes, "uses ten watts of power to perform 10^{16} synaptic events per second, whereas a state-of-the-art computer uses 100 watts to perform 10^{9} instructions per second. At this rate, a computer as powerful as the brain would burn 10^{9} watts: one gigawatt!"[2] This, by the way, is more than the peak output of most generating stations; in the present state of technology, you have the choice of powering half a million houses or one human intelligence.

At the same time, though, the brain is extremely slow: most neurons can fire no more than once every five milliseconds—a rate that means that the things we do over an interval of half a second (a notice-decide-act routine, like crossing the street to avoid a bore or drawing a gun in self-defense) can involve no more than one hundred consecutive calculations.[3] As a computer, my brain has a clock speed half a million times slower than my laptop—and that's *after* two cups of coffee.

How can this 10-watt, syrup-slow bag of cells still outperform any computer not linked to its own dedicated power plant? Through massive parallelism. Because each neuron connects to between one thousand and ten thousand others, computational tasks can spread widely through connected processing centers like a mood through a crowd. Where a classical computer functions linearly—start at the beginning, go on until you get to the end, then stop—all relevant parts of the brain work on parts of a problem simultaneously, passing back and forth provisional solutions based on partial data.

You're doing it now, reading this: each eye is currently looking at a different letter, leapfrogging down the sentence, while language centers in the brain make inspired guesses at what will come next.[4] This preference for speed over certainty is what makes human readers so much more capable than machine readers: we need very few clues to infer meaning. Take a well-known example from Graham Rawlinson of Sussex University: "You could ramdinose all the letetrs, keipeng the first two and last two the same, and reibadailty would hadrly be aftcfeed . . . The resaon for this is suerly that idnetiyfing coentnt by paarllel prseocsing speeds up regnicoiton. We only need the first and last two letetrs to spot chganes in meniang. This was not easy to type."[5] When seeking sense, we are surprisingly typo-blind, which probably explains why it is such a good idea to get someone else to do your proofreading.

Parallel processing introduces two further quirks into the brain's behavior that take it yet further from the simple picture of a general-purpose computer. The first is *modularity*: dividing the labor of thinking into parallel streams means that separate systems in different areas of the brain work on various aspects of a problem simultaneously. And if our brains can only do two hundred sequential calculations per second, it's clear that, while some of these calculations will compute basic sense-information like location or intensity of a stimulus, others will have to be dedicated to much higher-order matters. Take (again) reading: as you sped through Rawlinson's tangled sentences, you were processing at the same time, in different regions of your brain, many distinct varieties of data: from simple assessments of light and dark, edge and shape, to grammar, syntax, meaning, and memory—plus, no doubt, an extra element of bemused attention: "what *is* this?" Moreover, these distinct streams of thought must operate automatically: there is no executive function to dole out the work—because there isn't time to do that. We can only perceive the world in terms dictated by the brain's own procedures. No blank slate here—the human mind comes preformatted.

The second un-computer-like aspect of brain function is its *nonlinearity*. Taken at the level of the individual neuron, the task is straightforward: to fire or not to fire, on or off—which seems helpfully binary. But the precondition for a neuron's firing, the action potential, is collective: the sum of exciting and inhibiting inputs from neighboring neurons, including neurons which are themselves excited or inhibited by our particular neuron's past behavior. If the sum of influence passes a given threshold, the signal is given. The linkage is not fixed, as with the logic gates of a microcircuit, but probabilistic and recursive, as with a group of friends deciding whether to eat here or at that place down the street. This means that at the most basic level of brain function there is an element of the nonlinear: imperceptible changes may lead to widely divergent results, identical inputs to opposite outputs. As with the weather, there is a permanent germ of unpredictability in how we think. And since higher- and lower-level modules are working in parallel on the same perceptual problems, this unpredictability can set the stage for error at many points, from losing your keys to effusively greeting a total stranger at a party under the impression that she is an old friend.

It is also this unpredictability, of course, that *turns* strangers into old friends—that makes love so exciting and human company preferable to that of computers. Inputs don't map to outputs: we are not bound to do the same thing in the same circumstances any more than tomorrow's weather must be like today's. The nonlinearity of the brain's function means that, even though its local behavior may be completely determined by the laws of electrochemistry, our experience will still feel essentially, delightfully, like free will.

THE MAN WITH THREE BRAINS

There is a burglar in the house—you're sure of it. Those creaks and bumps are not accidental, and the children are asleep. The lights are all off. A chilly, liquid feeling begins to form in the pit of your stomach. What should you do?

What you should do, of course, is follow the emergency plan you probably never made: call 911 quietly on the cell phone you left by the bed, then, if possible, use the same escape route you practiced (didn't you?) in case of fire to get the family out; alternately, assemble them swiftly but calmly in the room you previously supplied with a strong lock and wait for the police. What you *actually* do is either a) think about screaming and running away; or b) think about going down there and *killing* the burglar. If there are two of you, it's likely one of you will favor screaming and the other killing. None of this, of course, is logical.

We are all aware of our capacity for irrational spasms of emotion. When some guy cuts in front of you on the freeway while leaning on his horn and giving you the finger, you would need the equanimity of a bodhisattva not to imagine his car exploding into a fireball. People do snap: according to one study, around 7 percent of Americans commit serious assaults because of *intermittent explosive disorder*—what a friend of ours describes in himself as "the red intolerables."[6]

It appears that some kinds of behavior go straight from perception to action without waiting to pass through the thinking brain at all—basic responses to danger or desire too urgent for consciousness. Sleep disorders can allow these old, dark urges to surface. In 1987 a Canadian man drove fourteen miles through the night, asleep, and killed his mother-in-law—with whom, it should be said, he had had entirely friendly relations during the day (she called him "the gentle giant").[7] In a case reported in 2004, a sleepwalking Australian woman would regularly leave the house to have sex with strangers (sleep sex, although only identified as a phenomenon ten years ago, is apparently not that rare and has been used as a successful defense in rape cases).[8] Lack of consciousness clearly does not mean lack of ability to function.

The source of the problem is also the source of many of our remarkable capacities: we are using three brains at once. Paul MacLean, the originator of the "triune brain" theory, described these as being like a pair of fists in thick woolen gloves, held up knuckle to knuckle.[9] The palms, deep down by the brain stem, are the original reptilian brain, very little different from that of our scaly ancestors in the Triassic. In reptiles and in monkeys, these structures control primal behavior: basic matters of animal identity, such as hunting, foraging, mating displays, preparing a home-site, flocking together, defending territory, bullying, and defecating in one chosen place—all, you will notice, things we do too.

The fingers of the fists are the mammalian cortex, a still-unconscious but adaptive region that allows mammals to do things reptiles generally can't: learn new behavior, care for families, develop social bonds. The subjective experience of the mammalian cortex at work is neither the blind reflex of the reptilian brain nor the fully formed idea of conscious thought, but rather the swirling dynamic of *emotion*—and if you have lived with a dog, you will know how powerful and subtle its play can be. Loneliness, sympathy, cantankerousness, mother-love—we probably feel these much in the way other mammals do, because we share the mechanism by which they feel.

Only at the woolly gloves, the neocortex, do we reach the special qualities of primate thinking—and only the thumbs, the prefrontal lobes, set us humans apart, with our self-examining consciousness and our attempts at self-regulation through logic and abstract example. Yet the thumbs cannot guide the hands in all they do, nor does our specifically human brain have full control of its mammalian and reptile forebears. We experience life at every level simultaneously.[10] Rage, fear, lust, and revulsion can sweep all wise counsel before them; melancholy, nostalgia, and pity undercut our resolution. When you ask yourself, "What am I *doing*?"—when you, say, take the drink you'd sworn not to, insult your best customer, or slink back into a doomed relationship—it may be because the mind you call "I" has been outvoted. On the other hand, logic without impulse and emotion can be equally dangerous: those for whom, through injury, the prefrontal lobes operate without the input of the emotions are poor at judging risk and fail to understand the motivation of other people.[11]

THE UNRELIABLE EYEWITNESS

Is seeing believing? Seeing was certainly the summit of science for Francis Bacon: as close to certainty as human experience allowed. Shakespeare's Othello echoed Bacon in demanding "ocular proof" of his wife's infidelity—and discovered, poor man, how far from certain such proof really is. Seeing *ought* to be believing: we devote enormous resources, more than a third of our brains, to our visual capacity.[12] The eyes themselves are formally part of the brain, with the cornea an adaptation of the protective dura mater and the retina simply a specialized patch of neurons extruded through small holes in the skull to meet the world face-on. Sight samples the most pervasive and revelatory aspect of our world: the play of light conveys so much information about our surroundings—

shape, color, texture, distance, movement—that we should be able to say of it, "Here is experience in the raw; I saw it, so it must be true."

I am a camera, then—but there's one problem: it is very difficult to get a camera to see as I do. Ask any cinematographer and you will be told the hardest part of the job is to simulate the natural, to control the depths of shadow, to retain lifelike colors across interior and sunlit scenes, to frame each shot in a way that makes the viewer forget directorial cleverness and concentrate on the story. As simple a shot as following a character from outdoors to indoors will require, if it is to appear normal and ordinary, four people serving the camera (operator, dolly grip, focus puller, iris puller) and corrective blue filters over every light fixture in the place. Making things look real is a highly synthetic business—the reason Hollywood movies are so expensive and cheap "reality" television looks so fake.

Not a camera, therefore, but a producer, an impresario. Vision isn't simply recording, or it would look as bad as home movies. It is, instead, a complex interplay between the world and our expectations, where what we *think* we should perceive trumps reality every time. Don't believe it? Here's a chance to see it.

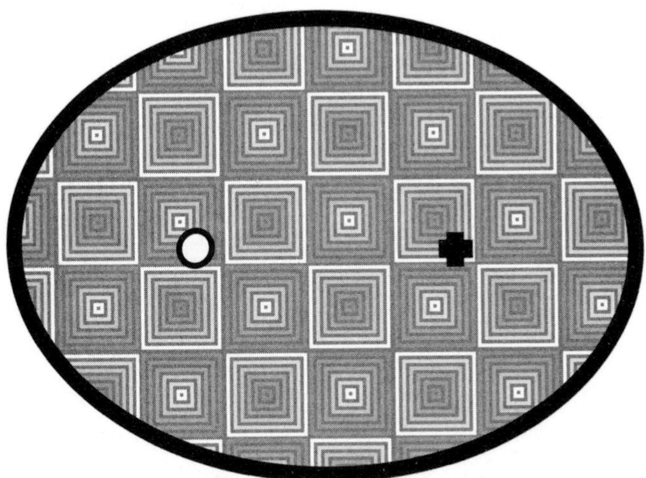

Close your left eye. Holding the book at a normal distance, concentrate your gaze on the white dot on the left. Then bring the book slowly closer, until . . . *hop!* The black cross has disappeared—gone completely. You've just mapped the location of your blind spot, the surprisingly large area of your retina where the optic nerve attaches and there are no light-receiving neurons. The interesting point in this experiment isn't that you have a blind spot (since you already knew that); it's that you filled in the missing information with a piece of the background pattern—you didn't see *nothing*; you saw something your brain made up. With practice, you can learn to do the same trick while you wait to be served in busy restaurants: by concentrating on, say, the end of your breadstick, you can make the head of the man at the next table disappear and watch as he shovels salad into empty space.

Of course, this fiction simply makes up for a defect in vision; it would be far more *un*real always to see a black spot floating around than simply to replace it with a generic background. As synthetic experience, it seems pretty mild.

Try this, then:

The figure on the left might look to you like a fortification plan or a Navajo blanket motif, but to *everyone* it's clearly a group of contained shapes, flat on the page. On the right, though, is a synthetic object that you can't help seeing, although it isn't there: the white triangle that seems to float slightly in front of the page. It even appears whiter than the space around it. It's as if the visual system gives it greater importance *because* it is inferred rather than sensed directly.

Similar inference turns our 2D vision into a 3D world.

We assume light comes from above; that's why the square on the left looks like a calculator button and the one on the right looks like the chalk for a pool cue. Turn the book upside down and the objects switch their identities, because we also assume that a source of light remains where it is when objects move. What we call illusion is really unshakable but unjustified assumption: our beliefs about the world control what we are *able* to see.

Why should we hobble ourselves in this way? Why can't we allow life itself to teach us what to look for? Why shouldn't we perceive each thing as it actually is, rather than run ourselves into the quicksand of illusion?

The answer comes down, again, to resources. Every second, each of our eyes supplies the rest of the brain with the same amount of data as an Ethernet connection: 10 megabits. This already represents a drastic slimming down of reality, only providing color vision for the center of the image and encoding as much information as possible through signals from metabolically cheaper, if less responsive, "sluggish" retinal cells.[13] But even this meager data stream would appear as brutally overwhelming as a fire hose if we were actually to be aware of it; the only way we can deal efficiently with experience is to throw out most of it and compare the rest with expectations.

Thus it appears (though some neuroscientists disagree) that we have two separate visual pathways in the brain with distinct, if related, goals.[14] The *ventral system* largely busies itself with the question "what is this?" Specialized neurons along this pathway respond to specific visual stimuli like edges, orientation, shadowing—even, remarkably, faces of our own species (the equivalent neuron in the macaque brain won't fire at the sight of even *very* hairy humans).[15] The critical task of this ventral pathway is categorization, so "I don't know" represents an undesirable failure of the system. The *dorsal system* handles "where is this?"—or perhaps more accurately, "what can I do with this?" It relates the body to its visual field, letting us send our hands and feet out into the territory our eyes have scouted. Both systems can be fooled, but they are fooled differently. Optical illusions, like the ones we have seen here, manipulate the expectations of the ventral system to make us misidentify one thing as another or see what is not to be seen. The dorsal system is generally not fooled by these, because it has no business with identification. Take this:

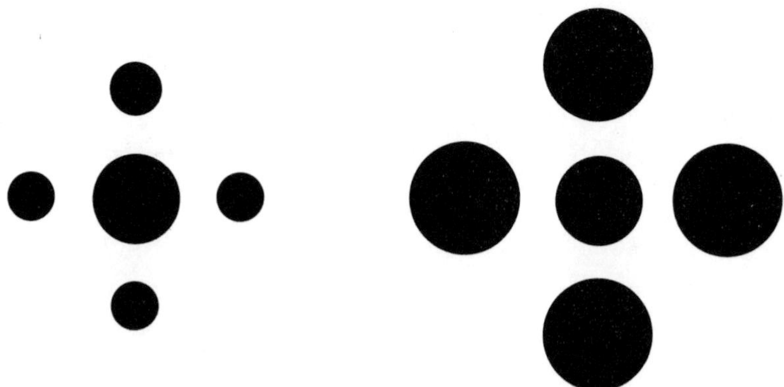

You may falsely see the central spot on the left as bigger than the one on the right, but you *reach* for them accurately, shaping your hand to the correct size. One patient with severe damage to the ventral system, who could no longer perceive or name shapes, could nevertheless still put them through the appropriately shaped hole.[16] The dorsal system's illusions are less specific but, for that reason, can be much more overwhelming.

"There was a flash and I was knocked backwards—except it wasn't me anymore. I could see myself lying there, the paramedics working to revive me. It was as if I was floating in the air, watching the whole scene. That's when I thought, 'This is it. I'm dying.'" The out-of-body experience has become a cliché of brushes with death (although it can also be brought on by fainting, drugs, damage to the temporoparietal junction, and at least at Laurentian University in Canada, powerful magnetic fields from a specially wired motorcycle helmet).[17] Recently, though, two groups in Stockholm and Lausanne have produced the same effect in healthy, fully conscious people using only a visual illusion: by fitting subjects with a pair of virtual-reality goggles and feeding them the view from a camera a few meters behind them, the experiment gave them a strong sense that their conscious selves existed outside their physical bodies.[18] When, for instance, a researcher punched the air near the camera, the subjects' galvanic skin response spiked in anticipation of pain—although they could see themselves sitting well out of harm's way.

What does this tell us about being wrong? Something fundamental: the brain is stubbornly unwilling to be flummoxed. When the senses provide unignorable information that seems to contradict reality's laws, we create a new version of reality that incorporates the information—no matter how far-fetched that version may be. "My conscious self has left my body to float in the air" is a much less *likely* explanation for a phenomenon than "my sensory apparatus has gone haywire," or the strictly appropriate "I have *no* idea what's going on." Nevertheless, it is the explanation immediately assumed by the experimental subjects—as well as by the hundreds of sincere, sober people who have returned from the dark tunnel leading to that white and kindly light.

Our visual errors differ from one another, therefore, not so much in their innate appearance as in their apparent meaning. Broadly, these phantasms separate into four families: *simple* illusions are usually matters of false identity between equally likely objects ("I thought I recognized you"); *significant* illusions are visions with personal importance ("a divine light appeared to me"); *psychotic* illusions suck us helplessly out of normality ("the world became full of horrible faces"); and then, there's the fourth and oddest category, *Bonnet* illusions, where failing eyesight populates the world with quaint and beguiling objects, such as tiny steam trains, tiling patterns or brickwork, little men in unusual hats, free-floating gargoyles.[19] The interesting thing about Bonnet syndrome is that sufferers never confuse the wild and dreamlike world in which their visual system has confined them with reality, nor even consider their hallucinations to have personal significance—they just "keep seeing things."

Thus, there is no real or illusory for us, but thinking makes it so: to save on expensive resources, the brain puts things in categories and assigns likely explanations to sense-experience long before it reaches the conscious decision-making mind. The world we think we see is actually an executive summary, helpfully condensed and annotated by unseen cerebral assistants.

AN ANIMATED WORLD

The evidence of creation, said the biologist J. B. S. Haldane, showed God's "inordinate fondness for beetles." The evidence of our illusions shows that we, too, are fundamentally naturalists, and that our brains assume we are still in the jungle.

Although our lives are increasingly dominated by the inanimate (cars, computers, cell phones, music players, microwave ovens), their designers work hard to imbue all this dumb stuff with subtly biological characteristics—giving them "happy" grilles, silky skin textures, chirpy alert tones, and softly yielding buttons or shift-sticks. This is because our brains respond differently to animals than to other objects. In a recent test, people shown pairs of photographs incorporating subtle changes were much better able to spot the difference when it involved an animal than when it involved a van, even though the van occupied much more of the picture.[20] We can infer animal—specifically, human—motion from remarkably few clues: a shifting pattern of white dots on a black background.[21] When these moving dots mark the position of the joints of a walking person, the observer spots it immediately—and can even specify the size, sex, and indeed *mood* of the person with surprising accuracy.[22] Within the brain, the sight of moving people or animals activates further areas beyond the visual system that are not activated by, say, falling rocks or swaying trees: memory, guesses about intention, emotions. So accurate and nuanced is this response that the computer-animation industry has largely given up trying to mimic human motion with software physics engines and instead makes cartoon characters credible through "motion capture" (filming an actor in a suit with white spots on it).

Cartoons also demonstrate how hardwired our assumptions about animate and inanimate motion are. Very small children find cartoons funny, which makes it clear that experience is not the basis of expectation—a three-year-old cannot have *learned* what should happen when a grand piano falls out of a skyscraper window or a coyote runs off a cliff, but her delight when the piano bounces, or the coyote stops in midair to wave forlornly before disappearing, reveals that her expectations are ready-formed. A three-year-old will also willingly accept the *illusion* of biological motion: take two blocks, one bigger than the other, and put them on the table. You and the child can pile them on top of each other and both of you will expect them to behave in a predictable, blocky way. But if you make the bigger one dart toward the smaller one and whisk the smaller one out of the way, and then ask, "what's happening?" the child is likely to say "the big one's chasing it," rather than the more realistic "you're just being silly."[23] From Charlie Chaplin's dancing dinner rolls to ventriloquists' dummies, we find a strong temptation to infer an animate being behind certain kinds of motion—and, as we saw, we can infer motion from very few clues. It's probably the reason people so often spot large animals in unlikely places: yetis have been seen not just in the Himalayan wilderness but in the suburbs of industrial Newcastle in England.[24] The right pattern of light and shade calls up the brain's useful predisposition to see beasts—and we become convinced not by what we've seen but by what we're wired to assume.

When a little priming is added to this predisposition, we start seeing the most unlikely things—sea monsters, big cats, Elvis, Lord Lucan—in the most unlikely places. In 1978, Rotterdam zoo reported the escape of one of its red pandas; hundreds of helpful people called in, having spotted it in places all over the Netherlands—when, in fact, it had been run over by a train just a few yards from the zoo fence.[25]

SAY WHAT?

Hearing is, in many ways, a better reporter of the real world than sight. It gives us a 360-degree impression of our surroundings, not the front-facing funnel of binocular vision. It senses a larger proportion of the frequencies produced in nature than the narrow, if lovely, gap between infrared and ultraviolet. Where sight cannot separate flashes closer than one twentieth of a second apart (or there would be no movies or television), hearing distinguishes individual clicks down to an interval of one thousandth of a second. Sound encodes the rich symbolic power of language and the emotional truths of music; it duplicates through tone of voice many of the clues to character and mood that we attempt to read from facial expression. This may be why becoming deaf seems more a banishment from life than blindness. Blind people can use the input from other senses to activate the brain's visual circuitry, automatically building up from sound, scent, and touch a mental impression that it would be no great stretch to call "a picture." Deafness, however, demands the conscious effort of attention, of *looking*, to supply the missing information. It is like having to remember to breathe.

Hearing may make shorter intuitive leaps than sight, but it too is subject to illusions. The most pleasant of these are "mondegreens," named by the author Sylvia Wright from her youthful mishearing of the Scottish ballad that actually says, "They hae slain the Earl o' Moray / and *they layd him on the green*"—not, alas, "the Lady Mondegreen."[26] Children, with their relaxed expectations for logic, are a rich source of these (pledging allegiance to "one Asian in the vestibule, with little tea and just rice for all"), but everyone has the talent to infer the ridiculous from the inaudible—and, what's more, believe in it.[27] Here, at least, we *do* behave like computers, in that our voice-recognition software has little regard for probability but boldly assumes we live in a world of surrealist poets. We are certain that Mick Jagger will never leave our pizza burning and that the Shadow knows what evil lurks in the hot cement.[28]

This snapping to certainty also appears in the phenomenon of "sine-wave speech," where a spoken or sung phrase is degraded down to a series of pops and clicks. They seem to be mere noise until you are told what the original phrase is—at which point it becomes impossible *not* to hear the sense through the static.[29] Coming to a conclusion is just that: it irreversibly terminates the guessing process. Once we have made sense out of our senses, we cannot go back. (This irreversibility can operate on a yet more abstract, if less exalted, plane: once you discover, for instance, that *The Rime of the Ancient Mariner* can be sung to the tune of "The Yellow Rose of Texas," the two will remain bound inextricably together in your mind—sorry about that).[30]

Vision can help us to recognize what our ears are actually hearing. We listen preferentially to what we can also see (one reason for the "cocktail party effect," where we can follow a conversation through a welter of similar noise that would defeat any microphone). Amazingly, the brain adjusts perception so that speedy light and slow sound appear to arrive in sync for things up to thirty meters away—a good horizon for coming to decisions about the unknown.[31] But vision can also fool the ear. In the *McGurk effect*, seeing someone, for instance, mouth the syllable "ga" while you hear the syllable "ba" produces a certainty that you have actually heard "da." The effect extends to whole sentences (one worrying example is a confusion between the phrases "he's got your boots" and "he's gonna shoot") and it's robust: even when you know what's happening, you can't turn it off. It goes a long way toward explaining the success of ventriloquists (and talking parrots, who have no lips to move) and why men with mustaches are harder to understand in person than they are on the telephone.[32]

When the senses help one another, they are doing so to the same end: if sight is on the lookout for animal—especially human—motion, hearing is on the earie for something similar. We respond, as every married householder knows, very differently to the sound of footsteps than to the tapping of a branch on the window. A recent study from the University of Lyon in France points out that vision and hearing share a motion-perception process in part of the brain called the posterior superior temporal sulcus, where the assumption of a biological source for sense phenomena seems to reside: a "who is it?" rather than a "what is it?" processor.[33] We do not scan our surroundings aimlessly; instead, like SETI's observers searching the universe for patterned transmissions, our focus is on detecting beings like us. All these biases in sense perception tend to purge experience of its randomness. Faced with puzzling information, we invent significance—inferring structure, meaning, and agency to resolve the mystery. Life as interpreted by our brains is something like a children's story: with more animate beings than there really are and a more prominent role for me, the brain owner, than experience would justify. And of course the most memorable children's stories are the frightening ones.

"I was sweating but cold and the feeling of depression was noticeable—but there was also something else. It was as though something was in the room with me. Then I became aware that I was being watched, and a figure slowly emerged to my left. It was indistinct and on the periphery of my vision, but it moved just as I would expect a person to. It was gray, and made no sound. The hair was standing up on the back of my neck—I was terrified."[34] A corrugated iron shed in Britain's industrial Midlands would not, you'd think, be the prime site for a haunting, nor would Vic Tandy, a "hard-nosed engineer," be the expected channel for otherworldly visitations. Yet there was undoubtedly something very odd going on in his medical equipment research lab: not just Tandy but other staff reported feelings of dread. Some found themselves speaking to distant or absent colleagues, thinking they were close by. Yet these paranormal effects were not constant: there were times the shed seemed no more possessed than any other shed.

Luckily for science, Tandy was not just an engineer; he was a competition fencer. The night after seeing the apparition, he happened to come in to repair his foil and clamped it to a bench near where he had most felt the shed's characteristic unease. He found the free end of the blade was vibrating wildly. But instead of taking this as further evidence of the occult, he deduced the existence of a low-frequency standing wave: infrasound. A thirty-foot tube, like the shed, would resonate at a frequency of around nineteen cycles per second; this is just too low for humans to hear, but it does produce physical effects: hyperventilation, muscle tension, anxiety, and—at sufficient amplitude—smeared vision from sympathetic jiggling of the eyeball. And the source of the air movement generating the infrasound? A ventilator fan. When this was turned off, the lab ceased to be an abode of nameless fear and returned to its native cozy decrepitude.

Nineteen cycles per second is a very, very low D sharp: a tone that can be produced by some church organs, adding a shudder of the unearthly to sacred music. Many thirty-foot chimneys, such as serve the guest bedrooms of old houses, will also resonate at this frequency as the wind sighs over them, a giant boy blowing across the mouth of a lofty bottle. Damp caves, dusty corridors, abandoned tunnels—all, in the right acoustic conditions, can share the occult power of Vic Tandy's shed, convincing our brains, always willing to assume an agent, that we are in the presence of something or someone beyond our ken.

Our brains make such mistakes because they play the percentages. It's possible that mammoths and mastodons may have communicated through infrasound, as do modern elephants. So it might have been useful for early humans to be able to imagine a threat when sensing low frequencies, even if they were not conscious of them. In general, we tend to associate anxiety with things we have only inferred, rather than sensed directly: these are the uncanny, the spooky, the haunting, the alien.

In a risky world, it's probably wise to assume that something hovering on the fringes of sense means us no good. It would be unwise to maintain doubt too long, questioning experience as the tiger creeps ever closer, so we compound our fear with certainty. It is better to make any decision than none, because it is better to be wrong than lunch. So it is not necessarily a sign of schizophrenia to see figures in the shadowy woods or hear voices in the babbling stream, nor indeed to identify those figures as ghosts of local legend and the voices as those of aliens (or, for earlier imaginations, the dances of satyrs to the song of naiads). This is just our senses doing their job—inferring significance—although somewhat too well. When the illusion is broken and we see the truth, the world loses a little meaning for us. We laugh as the tension loosens, but deep down we are slightly disappointed.

ATTENTION! THIS REALITY MAY BE EDITED

Coherent experience requires us to throw out most of our sense information: you were, probably, entirely unaware of your left elbow—until now. We cannot be mere sensory sponges or we would rapidly overload our mental resources. Most of our contact with the world must be active, choosing what we wish to perceive. It's why language distinguishes between *looking* and *seeing*, *listening* and *hearing*—and beyond both, *noticing*. At its best, this choice of attention marks the highest human achievement, where an unwillingness to be distracted produces work that is both complete and internally perfect. At its most prosaic, it makes us absentmindedly put the sugar in the fridge and the milk in the pantry, or drive to school instead of the supermarket on Sunday because there is something interesting on the radio. At its worst, it makes us die or kill unnecessarily.

Attention starts at the simplest sensory level: our eyes do not sweep the room radarlike to take it all in, but shift from focus to focus in a series of jerks called *saccades*. Here, filmmaking has it right: the eye may not be a camera, but visual experience *is* an edited sequence made up of cuts. Film editors have long known this. For any scene, there is a natural way to cut it where the audience will be completely unaware of the editing and simply find that significant elements (her face, his eyes, the gun on the table) are presented in the order and with the timing that human attention finds most comfortable. This technique naturally involves leaving a lot of footage on the floor: between saccades and blinking, the human visual system is actually switched off for up to a quarter of the time, during which a lot can happen.

The brain has several tricks to make up for this incomplete record. The first is to fill gaps in time, as it fills in the blind spot, with the most likely placeholder. If you shift your gaze to a clock with a moving second hand, you will see it appear to "hang" momentarily, with the first second seeming about 10 percent longer than the next. This is because your brain has filled in the roughly one hundred milliseconds of the saccade with what you saw *after* it; your sense of "now" actually runs at a tenth-of-a-second delay (which is why any reaction faster than this counts as a false start in track events).

The brain also simplifies experience by bundling. See three birds and you can track them individually; see five and they are on their way to becoming a flock. We can prime our attention to look out for things as distinct as a particular face or as vague as "colors that would look good in the living room," but we can only keep around seven similar things in mind at once; more than that and things become categories—or they become totally confusing. This is the secret of prey animals—the swirling "bait ball" of herring, the plunging herd of wildebeest: by sheer numbers they overwhelm the predator's ability to pay attention to any one potential victim.

The essence of attention, though, is extracting the significance from a scene, and that usually means looking for the unexpected. Traditional training for film editors used to go like this: "Look at the shot and snap your fingers when you've seen enough of it" (almost all snap their fingers at nearly the same moment). "Now say what you want to see next: her reaction to his line? The thing he talked about? The view in through that prominent window in the background?" Any image forms expectations; attention responds to novelty. We *look* at what has surprised us in what we see. Babies, for all their myopic and distracted air, are doing this all the time, responding attentively to faces, moving objects, and most of all, the results of their own expectations. Seven-month-old infants, having seen one doll and then another added to it behind a screen, react strongly when the screen is removed to reveal only one doll.[35] Things that go against expectation are not just random. They make us pay attention; they must *mean* something.

Macaques do the same thing; in their inferotemporal region (the highest level of purely visual interpretation), they have distinct neural encoding for things they have seen before and things they have not.[36] This encoding occurs in a relatively small number of neurons (several hundred) and is very quick; the readout of identification takes only about thirteen milliseconds—far shorter than a saccade, which means the monkey will know that something is unfamiliar *before* it sees it. Less advanced creatures control gaze direction through the superior colliculi (an evolutionarily ancient structure, looking, in humans, like a bracket fungus on the brain stem), and we often do too, latching onto pattern-breaks in as unconscious a way as did prehistoric reptiles. In general, the speed and dispersion of attention mechanisms in the brain mean that, much as we may feel that we are consciously deciding what is important in our world and choosing to focus on that, the decision has largely been made before we are aware of it. "Look here upon this picture, and on this," says the brain: it's hardly surprising, therefore, that we find it so hard to believe in coincidence and tend to infer cause and effect without much evidence. We are never actually seeing the raw data.

This is not to say that we are the permanent dupes of our perceptual systems. We can and do enforce our wills, stubbornly choosing to seek out what we think will be important. The results can be remarkable: Ted Williams trained himself to see a baseball *as he hit it*, to such accuracy that he could tell if his bat had struck across one seam, two, or none (this is claimed, in the current state of medical knowledge, to be impossible, but it was witnessed in action by a reliable and disinterested umpire).[37] Adepts at kendo, Japanese fencing, practice the discipline called "looking at the mountain": maintaining a wide, unprejudiced visual attention—since an incipient attack may signal itself as much in the twitch of a toe as the blink of an eye. A good financial trader learns to read significant patterns and their exceptions in what, to the rest of us, seems the meaningless clutter of his Bloomberg screen. We admit and admire the power of concentration, which is why we say, "look, and thou shalt see," not the other way around.

Really to look at one thing, though, is to become blind to everything else. The most famous evidence for this is the experiment in which subjects were asked to watch a video of a basketball game and count the number of passes made by one of the teams but not the other. Concentrating on this task, the majority of viewers simply did not see the person dressed in a gorilla suit who walked across the court, paused in front of the camera, and moved on.[38] This may seem a slightly artificial task, but it shows how little we notice that which does not concern us. As William James is said to have explained, absentmindedness is simply being present-minded somewhere else—we regularly send our minds off on errands, or allow others to do so for us.

Ask anyone, "How many animals of each kind did Moses take on the ark?" and the answer you'll get is "two"—not "*Moses?*" Our conscious attention is task-driven and we are remarkably docile about having the task set for us. Thus a clever adolescent will ask his mother,

"Should I wear my jeans or my camo pants *when I go out tonight*?" just as a clever salesperson will say, "You can have the dividends *from your new investment* sent by check or deposited directly." In more serious situations, this tendency to grasp only part of what's going on can have its uses. Much of the point of military training is to occupy enough of the mind with following correct procedures for it to become blind to the fact that someone is trying to kill it: in the traditional phrase, to make the sergeant more frightening than the enemy. In the courtroom, where every tight-knit case threatens to unravel into a tangle of confusing detail, the advocate's crucial job is to direct attention, to make one comprehensible feature, no matter how logically irrelevant, stand in the jury's minds for the whole: "If it doesn't fit, you must acquit."[39]

Seeing only what we consider important is efficient, but also risky. On January 8, 1982, the U.S. Air Force Thunderbirds display team was training over Area 65 at Indian Springs Field in Nevada. The four T-38A Talons were practicing a loop in the tight diamond formation for which the team is famous. It's still not clear exactly what happened to the lead plane; somehow there was insufficient back-pressure on the control stick to complete the loop. What *is* clear is that there was nothing wrong with the three other planes, but the intensity of their focus, summed up in the Thunderbird catchphrase "I follow my #1," meant that the diamond flew in perfect formation straight into the ground.

Ignoring what we take to be extraneous information can be just as deadly: during the invasion of Iraq, two A-10 pilots from the Idaho Air National Guard were spotting artillery targets for Marine units on the ground, when suddenly:[40]

13:37.54 pilot POPOFF 36	Hey, dude.
13:37.56 pilot POPOFF 36	I got a four ship of vehicles that are evenly spaced along a road going north.
13:38.04 pilot POPOFF 36	. . . along that canal, right there. Coming up just south of the village.

There was only one problem:

13:38.49 pilot POPOFF 36	They look like they have orange panels on, though.

Orange panels were a sign of friendly forces. The Marine air controllers had said there were no friendlies in the specific area the two POPOFF pilots had previously been spotting; this was taken as a blanket description: "He told me, he told me there's nobody north of here. No friendlies." Within twenty seconds, "orange panels" turned to "something orange." In another forty seconds, POPOFF 36 said, "I think they're rocket launchers." Twenty seconds after that, he claimed, "They got orange rockets on them," followed another twenty seconds later by: "I think killing these damn rocket launchers, it would be great." A minute and a half later, POPOFF 36 strafed a patrol of British reconnaissance vehicles, killing Lance Corporal Matty Hull. But even as he completed his run, he asked the other pilot, "That's what you think they are, right?" He watched the other pilot line up for his own strafing pass, then asked again:

13:43.24 pilot POPOFF 36	It looks like he is hauling ass. Ha ha. Is that what you think they are?

It's as if something were still niggling at the back of his mind, although all his conscious attention was taken by the attack; indeed, he then prepared for a third run himself. Meanwhile down below, Trooper Christopher Finney, having rescued his wounded comrade from a burning vehicle, climbed back into it to radio that the patrol was under friendly fire; he was himself shot in the back and legs during the second pass. The news reached the pilots a minute later.

13:44.21 Control Lightning 34	Hey, POPOFF 34, abort your mission. You got a, looks we might have a blue on blue situation.

13:44.25 pilot POPOFF 35 Fuck. God bless it.

13:44.36 control Manila 34 POPOFF 34.

13:44.35 pilot POPOFF 35 Fuck, fuck, fuck.

POPOFF 36 is weeping as the tape comes to an end. In four minutes over Basra, he had discovered how closely distraction fits its old definition—madness.

AH, YES, I REMEMBER IT WELL

> When I would awake in the middle of the night, not knowing where I was, I could not be sure at first *who* I was; I had only, in its primary simplicity, a feeling of existence, such as might tremble in the depths of an animal's consciousness; I was more exposed than any cave-dweller; but then memory, not yet of the place in which I was, but of various other places where I had lived, and might now very possibly be, would come, like a life-line let down from above, to draw me up out of the abyss of nothingness, from which I could never have escaped by myself.

The most complex essay on memory in Western culture, Proust's *À la recherche du temps perdu*, offers two strong and closely related images of memory: as the divine rope and as a casting of precious objects:

> Rare as they became, those moments did not occur in vain. Through memory, Swann reassembled their fragments, abolished the intervals between them, cast, as in gold, an Odette of kindness and calm, for whom he was to make, later on, sacrifices that the other Odette would never have won from him.

Like all good spoken music, these words set up a sympathetic ringing from distant corners of the reader's own memory—Jacob's ladder, umbilical cords, the golden calf, Minos's daughter—all chiming to the same tune: memory is our most precious possession, our human birthright, the mark of significance that sets our lives apart from those of beasts. Of course, this is being unfair to the beasts. Even bacteria have some kind of memory or they would not be able to move from a nutrient-poor region to a nutrient-rich one—comparison over time *is* memory. Sea slugs can remember an unpleasant stimulus and avoid it by pulling in their gills. The stronger the stimulus, the more long-lasting the recall.[41] It may not have the incantatory magic of Proust's madeleine dipped in lime-flower tisane, but it is still memory.

What does set human memory apart is its complexity and its illogicality. If you were designing it from scratch, you would probably do as computer scientists do: specify a memory that combines stability, capacity, and speed of access. Yet our greatest mental gift, the internal culture we always carry with us, fulfills none of those requirements.

Take stability: you would think the essential requirement for memory is to retain a true record of the past. Certainly, this is what is expected in a court of law—"the whole truth." Yet, even when our or another's freedom isn't on the line, people remember falsely—and not just details but central facts.

The Mass-Observation movement of the 1930s believed that only by wide unprejudiced collection of near-random facts could the secret dynamics of social science reveal themselves. It encouraged diary writing, and as a result, several study participants kept daily records of exactly what happened to them in World War II, specifically during the London Blitz. Years later, two of the movement's founders, Tom Harrisson and Philip Ziegler, published some of these diaries and thought it might be interesting to go back to their surviving writers and compare memory with testimony. Ziegler reports:

> Memory plays fearful tricks . . . We wrote to the authors, asking them, without referring to anyone or any document, to tell the story again. Half a dozen obliged. Any resemblance between the original

> story and the version recounted 30 years later was almost entirely coincidental. They got everything wrong: time, place, sequence of events. In almost every case they moved themselves closer to the centre of events; what had happened to a neighbor now happened to them.[42]

These were not eager self-publicists hoping to appear on talk shows and secure lucrative book deals; they were, without exception, sweet little old ladies of London Town, pulled out of uneventful retirement to answer unwanted questions. Yet they could not be trusted on the most important event in their lives.

Little old ladies are not the worst culprits. In one famous study by Michelle Leichtman and Stephen Ceci, three- to six-year-old children were introduced to an actor playing a character named Sam Stone, who would briefly visit their class or kindergarten, without interacting individually with any of the children.[43] Later, the researchers would show them a torn book and a soiled teddy bear and ask if they knew who had done it. Some of the children had been primed over the previous weeks with occasional stories about Sam Stone, presenting him as a bumbling, accident-prone character. Others were questioned about the book and bear in a leading way, suggesting that people expected Sam Stone might be at fault. Some children received both these prompts, and among them, 72 percent of the preschool-age children said Sam Stone had definitely committed both misdeeds and 44 percent said they had actually seen him do it. One helpful three-year-old explained: "He played with one of the toys, and [we] said 'be careful!' but he didn't be careful with the toys." In a worrying side study, an audience of 119 experts in children's legal testimony shown videotapes of the interviews could not tell the difference between true and false recall, and perhaps unsurprisingly, rated the most inventively fictive child as the most credible—because her story was lively and detailed.

As Leichtman and Ceci explain, many factors can influence suggestibility: desire to please a questioner, desire for reward, fear of embarrassment. But they suggest that another mechanism is also at work, most obviously among the younger children. In imagining the vivid story of Sam Stone, Force of Nature—tossing books in the air, painting teddy bears with ice cream—they were combining two mental pictures, one true and one false, in a way that would be very difficult to separate when either was recalled. The true picture was dull and forgettable; the false one was exciting, and clearly important to the adult questioners: it had meaning, not just in the past, but also for the present.

All these mechanisms were plainly at work during the sudden flurry of mass-testimony child abuse cases in the 1980s and 1990s. In one instance, Margaret Kelly Michaels of the Wee Care nursery in Maplewood, New Jersey, was sentenced in 1988 to forty-seven years in prison on 115 counts of child abuse, the form of which—involving peanut butter, Lego blocks, and nude piano-playing—would seem an obvious figment of childish imagination if it were not for the further horrific details the state's investigators extracted from the witnesses, using techniques of suggestion so grossly tendentious that the prosecution was dropped on appeal five years later.[44] Many other day care workers, however, remain in jail, like the Texas couple accused of crimes (burying children with dead animals, painting in blood with bones, infant sacrifice) that appear, in the complete absence of any forensic evidence, like the result of too-early exposure to the works of Stephen King.[45]

The problem for a jury is that it has only two choices: it must decide if a child is lying or telling the truth. But memory is a tool of the present, not just a record of the past. The children in the Michaels case were grilled repeatedly by the investigators and their parents; they then testified before a grand jury in a large courtroom. Before their appearance at the trial, they were visited again by the prosecution team to make sure they had their testimony right. They then delivered it by closed-circuit TV from the judge's chambers—several, indeed, while sitting on the judge's lap. It would be an inhumanly strong-minded five-year-old who could keep the dull, long-ago truth intact against all this pressure.

If you want to watch a false memory being born, make an effort to imagine, vividly, your mother telling you this story: "Do you remember when you were about five and we were out shopping and you wandered off? And then that old man found you and you were frightened of him, but it turned out he was really nice and he brought you back to me? Do you remember what he was wearing and what I said to you?" With the usual few exceptions, almost everyone will find some tickle of familiarity about this story, an effect exploited in the lab of Elizabeth Loftus at the University of California, Irvine.[46] The "lost while shopping" story is one of hers, and subjects will happily supply the details: "The old man was wearing a flannel shirt"; "Mom said, 'Never do that again.'" They will do this even if told the story merely by an elder sibling. Loftus can also convince people that they really performed during an experiment actions they had only *imagined* performing (including such hard-to-forget matters as "rub the chalk on your head" and "kiss a plastic frog") or that they went to Disneyland and met Bugs Bunny (who is not, after all, a Disney character). She primed Alan Alda, visiting to report on her work for a PBS series, to believe that he loathed hard-boiled eggs because (as he had been led to remember falsely) he had become sick after eating them as a boy.

Why is this? Why do people provide such rich descriptions of things that never happened? It appears that several things are going on simultaneously. First, if an authoritative or credible source provides information, we tend to assimilate the information into our personal memories rather than holding it, as logic would suggest, in some separate mental category. Imagining oneself in a situation uses brain processes perilously close to remembering it. It's an effect that police and prosecutors exploit, consciously or unconsciously, as they run through testimony with eyewitnesses during the many months between a crime and a trial. We make use of the same effect to justify our own beliefs: wishful thinking will, over time, attribute a more authoritative remembered source for our opinions than is true. *Consumer Reports* said I have an excellent car, not the guy at the dealership; I read about this fact in *Nature*, not the *National Enquirer* in the checkout line.[47]

Second, holding memories is a dynamic process. We simultaneously build up and polish them, like pearls or gallstones. At each recall we discard details that seem irrelevant and, yes, add elements that would be appropriate.[48] If you are putting in the effort to remember that time you were lost while shopping, other memories can combine with a sense of likelihood to fill in the details, as the eye fills in a likely background: a kindly old man *would* wear comforting flannel; any mother would say, "Never do that again."

Finally, memory is dispersed through the brain, not stored in some central server. Some elements are amazingly distinct and local: experimental subjects have revealed in brain scans individual neurons that fire only on seeing the face or name of Halle Berry (with another for Jennifer Aniston, and another for the Sydney Opera House, and so on).[49] Visual components of memory reside in the visual system; verbal memories activate the areas associated with spoken or written language. The moment of remembering is a reassembly of experience from dispersed bins of sensory and abstract information; you can feel it in the different ways you call up the past depending on the task. "What can you tell me about honesty?" will start this reassembly from a different point than "what does the smell of apples remind you of?" Yet each eventually yields memories that are both vivid and significant—more vivid and significant, rich and meaningful, than life as experienced usually is. And where faulty recall retrieves only one element—information without sense experience, emotion without content—the brain has a fund of material to make good the deficiency.

Thus we "remember with advantages," in Shakespeare's phrase, gradually moving ourselves to the center of events and making our past a stately progress of logical, inevitable steps, in which the randomness of real events becomes a coherent and motivated story—happy or sad, but definitely meaningful. Kenneth Lanning, an FBI investigator of the “recovered memory” phenomenon (in which adults, aided by keen if credulous psychologists, dredge up instances of childhood abuse of which they had been previously unaware), points out that almost all these cases begin with the subject’s desire for help with personal problems and failures: drug abuse, lapsed relationships, self-mutilation, eating disorders.[50] For many, the desire to assign a specific—even if dreadful—reason to what is otherwise a pervasive meaningless misery can be just as strong as the desire in others to rewrite the past as a rosy and comforting myth. The same facts, without any conscious desire to deceive, can produce radically different interpretations in different people’s memories—the reason so many family Thanksgivings degenerate into theater of the absurd.

Why, then, is our memory so poor? The answer is, it isn’t, except in the role we have artificially assigned to it. The fundamental fact behind all this fiction is that memory has a purpose, but that purpose is not to remember accurately—it is to envision and plan the future. We consult it, not as we do a diary, but as we do an almanac—so we should not be surprised that it has about the same degree of trustworthiness.

With maturity, the proportion of white matter—that is, insulation—in our brains increases.[51] The electrical fizzing of childhood settles down into discrete consistent circuits. The experience we elaborate in memory shifts from the rapturous intensity of sense impressions (children have far more sensory neurons than adults, so, yes, chocolate cake *did* taste better when you were a kid) to the proverbs, parables, and Dreadful Warnings by which we adults justify our conduct. In routine life, we eventually acquire more than enough of these examples and find we need no more to guide us—the process of building new memory slows. Thus, as the scholar George Lyttelton pointed out in his own seventies, the mark of senescence is not what a man forgets, but what he remembers: the same old jokes, the repeated anecdotes of river-smooth delivery that set younger eyes rolling around the table.[52] We recall every significant detail, except how often we have told this story to the same person.

Twice as many people fear old age for its loss of memory as they do the physical incapacities it brings, but what they should *really* fear is the loss of memory's purpose: engaging with new experience. To keep hold of the divine lifeline, to remain at the bellows, smelting the precious images, we must put ourselves in situations where the old responses will not do. We should travel, meet new people, volunteer: go where life regains its youthful novelty and intensity—to which, we hope, we will add the wisdom of experience.

SEPARATED BY A COMMON LANGUAGE

PRESIDENT: Boy, you know, you never know.[53]

Those who lived through the Watergate scandal will recall the nation's shock when the first transcripts of the White House tapes were released. The copious use of "[*expletive deleted*]" first captured the popular imagination, but more troubling was the blurry mental focus in the Oval Office:

PRESIDENT: It isn't worth it. It isn't worth it. Damn it, it isn't worth—the hell with it.

What is the situation on your, uh, on the, on the little red box? Did they find what the hell that, that is? Have they found the box yet?[54]

The masters of the Free World were talking as if they were stoned; they seemed less Machiavellian than comatose.

Now, though, the Nixon Library has released some of the actual recordings—and the impression is entirely different. Through the hiss and buzz of the Dictabelt, we hear a drama made up of significant silences and unfinished attempts, as sparse and ominous as a Bergman script. Here, the president's thrumming, urgent basso intercuts with the astronaut-blandness of H. R. Haldeman:

PRESIDENT: Oh well, on that we'll destroy him.
HALDEMAN: And ah
PRESIDENT: It's his word against
HALDEMAN: You know from the big
PRESIDENT: His word against
HALDEMAN: Something like that
PRESIDENT: His word—what?[55]

As words, formally meaningless—but as *sounds*, weighty with cloaked menace. Later, Chuck Colson's church-supper folksiness barely conceals his feral determination to survive:

COLSON:. . . uh, uh, I don't want to get in the position of knowing something that I don't now know for the reason that, I wanna be perfectly free to help ya, and the only way I can help Ya [*sic*] and the only way I can help ya is, uh to remain as completely uh, you know, unknowing as I am.
HUNT: I see.[56]

The murky picture clears: what these men were really talking about was the thing they weren't talking about; they were sketching in negative the huge, troubling void in their midst. They garbled and stuttered as a sign that there was nothing coherent they *could* say—other than the unspeakable truth.

Presidents are not alone in incoherence. One's own speech, transcribed, rings as false as the sound of one's recorded voice: both seem so hollow, emptied of their familiar warmth and resonance. Those hesitations, repetitions, stopgap phrases—how can they possibly convey thought? What could I mean by "*I mean*"? What did I assume when I said "*you know*"? Our words are weak bearers of our thoughts, just as our senses are unreliable reporters of the world around us.

Travel guides are full of warnings about errors of gesture or custom (don't, for instance, make the finger-thumb circle sign for "OK" in Brazil unless you weary of life), but words are the quicker path to confusion. Flapping your arms and clucking in a Tibetan restaurant will at least get you what you want, but mentioning in France that you are *embarrassé* or *plein* will gain you only raised eyebrows (unless you actually *are* pregnant, in which case you should have used the feminine adjectival form). Nor does sharing a language necessarily make communication easier—remember this gem from State Department spokesman Robert McCloskey, put on the spot during a Vietnam-era briefing: "I know that you believe that you understood what you think I said, but I am not sure you realize that what you heard is not what I meant."[57]

If speech is just a conveyance for meaning—if we always have something to say before we say it—why do we talk to ourselves? A recent study from UCLA gives a clue: when people are faced with a stressful situation, *along with a word that describes it*, stress-boosting activity in their ever-nervous amygdalae dampens down.[58] As Matthew Lieberman, the lead author of the study, says: "Expressing your feelings in words short-circuits the body's reaction by preventing stress hormones from being released." The cries and whispers of the Nixon White House, therefore, were not simply attempts to communicate the unspeakable; they were also a way to fix and thus diminish the fears of the embattled staffers.

But only speech, with its innate music, can carry both messages—the calming terminology and the emotional subtext that is the real heart of the story. Written language struggles to match the subtle immediacy of the voice. As telephone conversation gives way to e-mails, texts, and instant messaging, we find our quick, bald words becoming less and less trustworthy. A recent study from New York University and the University of Chicago shows that people cannot actually tell the difference between sincerity and sarcasm in the e-mails they read—yet they presume that they themselves make this difference clear in the ones they write.[59] Two other researchers in the field, who had never met in person, found themselves shouting at each other after an escalation of e-mail misunderstandings. "It became very embarrassing later," admits one, "but we realized that we couldn't blame each other for yelling about it because that's what we were studying."[60]

Equally trivial blunders can have much larger effects: in 1893, confusion over Admiral Tryon's signal that the two divisions of his fleet should turn inward "preserving the order of the fleet" caused his leading battleships to collide, killing 358 men, including the admiral. It can be argued that the disastrous charge of the Light Brigade hinged on an injudicious use of the word "there." James Hanratty, the last man to be executed in Britain, was hanged on the ambiguity of his shout to his accomplice: *"Let him have it."* Was it, *"Let the policeman have your gun,"* or *"Shoot the policeman"*?

We call computer code "language," but only a computer would actually treat language as a code and assume that all meaning is contained within it; human communication is based far less on formal content than on rapport. The mirrored gestures, the synchronized smiles, the meshing of tones of voice all have one purpose: to establish a shared interest in mutual understanding for its own sake, an earnest of respect for *whatever* the other might think is significant. It is much easier to negotiate on the telephone with someone you have met face-to-face; going up one level of abstraction, Michael Morris at the Columbia Business School found that people charged with reaching complex agreements by e-mail did so much more quickly and effectively if they had first had a brief telephone conversation on some entirely unrelated topic: they dealt with each other thereafter as human beings.[61] As anyone knows who has bobbed and grinned through an evening in an unknown language, we hunger above all to be understood. It is not merely convenient; it is essential.

Clumsy e-mails disappoint this hunger; good poetry, on the other hand, piques, tempts, and satisfies it. In the widower Thomas Hardy's love lines to his late wife,

Numb as a vane that cankers on its point
True to the wind that kissed ere canker came[62]

we veer dangerously close to the point of incomprehension, yet the creaking, sibilant sound and the lonely intimacy of image bring us back, giving us a genuine rapport with this long-dead, far-distant poet. We feel that we understand, and oddly but inescapably, that we are understood—in our finer selves, our chambered emotions.

Philip Davis of the University of Liverpool has long been interested in this electric shimmer that accompanies the comprehension of an obscure but well-turned phrase—particularly in the works of Shakespeare, the eternally daring boy who climbs far out along the boughs of grammar to perform his capers among its thinnest twigs:

> I had a specific intuition—about Shakespeare: that the very shapes of Shakespeare's lines and sentences somehow had a dramatic effect at deep levels in my mind. For example—Macbeth at the end of his tether:
>
> *And that which should accompany old age,*
> *As honour, love, obedience, troops of friends,*
> *I must not look to have, but in their stead*
> *Curses, not loud but deep, mouth-honour, breath*
> *Which the poor heart would fain deny and dare not.*
>
> I'll say no more than this: it simply would not be the same, would it, if Shakespeare had written it out more straightforwardly.[63]

Davis and other colleagues set up an experiment to see what happens in the brain in these moments of grammatical ambiguity but semantic revelation. They chose a favorite trope of Shakespeare's, the dragooning of nouns into service as verbs—something that works wonderfully for him ("I could *out-tongue* your griefs") if less well in modern usage ("she could *podium* again for sure"). Subjects were given an electroencephalogram (EEG) scan, during which they read sentences that fell into four categories:

1) grammatically and semantically unexceptional: "I said you would *accompany* me";
2) grammatically and semantically incorrect: "I said you would *charcoal* me";
3) grammatically fine but semantically wrong: "I said you would *incubate* me";
4) Shakespearian—that is, grammatically unexpected but semantically valid: "I said you would *companion* me."

What Davis and his colleagues found was a phenomenon that neuroscientists and linguists have known about for years but have somehow kept secret from the English Department: the P600 and N400 effects. These are modulations of brain wave patterns in response to verbal cues. The P600, as the name suggests, is a positive modulation that appears around 600 milliseconds after a word that seems to violate grammatical rules; the N400 is a negative modulation, coming 400 milliseconds after something that defies comprehension. "I said you would charcoal me" sets both responses going; "I said you would accompany me" neither. The experience of Shakespeare's shift in word roles, though, is a regular tickling of the P600 with no corresponding N400: it sets up a tremor between "huh?" and "ah!" as the brain integrates conflicting views of the sentence. Nothing is laid out for us: we make the sense for ourselves, just as we made the missing triangle in the optical illusion—and for the same reason, this sense glows all the more brightly.

Such active, lively modes of perception seem to give us pleasure. Good jokes work on the same principle, breaking the local rules in the interest of a wider, skewed meaning. ("The sign said We Serve Breakfast Anytime," murmurs Steven Wright. "So I ordered French toast during the Renaissance.") Puns, even—yea, especially—Shakespeare's, may well fall so flat because they are neither unexpected enough to engage the P600 nor sufficiently meaningful to suppress the N400. Music can also set the P600-N400 dance going, both when heard in combination with words (the word "breadth," for instance, apparently becomes much more meaningful when combined with Strauss's *Salome* than with a piece of accordion music) and according to its own semantic structures of repetition and transformation, tension and resolution.[64] Well-behaved Salieri, say, should hardly trouble either wave form, but Mozart constantly challenges the expectations of an educated ear: "Can he *do* that?" "I don't know, but he *did*!"

Even visual images can create expectations of significance revealed through brain waves—a phenomenon so far explored by science only in a simplistic "what's wrong with this picture?" style, but perhaps a clue to why the same topics (Annunciation, Crucifixion, and Resurrection; the judgment of Paris, the pursuit of Daphne) resonate anew in each skilled treatment. Iconography is a kind of grammar; and the best painters know well how to make new sense of old forms. In all the arts and sciences, this almost physical pleasure of discovery, of finding meaning in unexpected places, is what Thomas Hobbes said distinguished us from all other animals: we have "a lust of the mind that, by a perseverance of delight in the continuall and indefatigable generation of Knowledge, exceedeth the short vehemence of any carnall Pleasure."[65]

We ought not to use such infant science as a club to beat up other, older forms of thought, but this aspect of neurology does provide a differently useful tool: a skimmer to separate the good from the bad in art. The surprising resolution of semantic tension, the intelligent messing around with expectation, is a mark of creative quality, and the more levels on which this happens (formal, sensual, literary, popular, spiritual, and so on), the more pleasurably engaged our minds will be—as long as we do our part by bringing a prepared mind. Works that manage this trick have the essential quality of being unimaginable in prospect but inevitable in retrospect. Thus, Warhol is good, but Rembrandt is better; the *Eroica* will always mean more than "My Ding-a-Ling," and a well-executed squeeze play is, on the whole, more satisfying than an upper-deck home run. So there, cultural relativism!

NATURE'S DRUGSTORE

"Information," insisted Norbert Wiener, founder of cybernetics, "is information—not matter or energy."[66] That is, we should not confuse the data being computed with the reality it may represent. This is certainly true, and it would be a shame if the brain's complexity were to distract us from the fact that, *whatever* is happening in there, it is all still information processing, with binary messages passed from cell to cell in patterns that encode larger significance.

That said, the brain's complexity means it has so far been impossible to build a formal model that can bridge the gap between the computational role of the individual cell and the actual experience of thinking and feeling. As we've seen, the larger part of the brain's functional apparatus is given over to unconscious processes for which, in the absence of formal symbolic language, we are reduced to using common words, with all their slippery vagueness. Moreover, the fundamental state of the brain—our alertness, moods, and basic mental posture—seems less a function of the information we receive and process and more a function *precisely* of matter: specifically, of the pharmacopoeia of endogenous drugs with which the brain habitually doses itself.

We're used to thinking of the brain as having a *natural* state, from which drugs detract in one way or another, nudging us away from an ideal neutrality. Thus, we ban opiates and hallucinogens; some shun alcohol, others caffeine. Yet psychopharmacology increasingly reveals that our apparent normality is itself a function of constant self-medication with powerful neurotransmitters. Romantic love may seem entirely natural, but romantic love is also marked by a massive increase in serotonin, similar to the effects of Prozac. Well, then, married love, or the love of a long-term stable relationship—what could be more natural than that? Yet simply block their oxytocin receptors and Darby and Joan will act like strangers to each other. Want to be a better father? Generate prolactin, as mothers do. Want to achieve enlightenment? Take up strenuous exercise and let anandamide carry you to runner's nirvana. Want to be more perceptive and engaged in the world? Stop eating: ghrelin, the resulting hormone, boosts learning capacity to such a degree that the researcher who found the connection speculated that "perhaps the cognitive brain is a side-effect of hunger."[67] If we assume, as we must, that thinking is defined as *what the brain does*—and what the brain does, at this level, is chemistry—then those who spurn psychotropic drugs as "unnatural" are in the difficult position of the Brahman given the microscope who found that his strict vegetarian diet was teeming with minuscule reincarnation.[68]

There is, though, an important distinction between *kinds* of drugs—whether natural or artificial—that defines both their action on the brain and the intrinsic dangers they may pose to a healthy mental life. The opioids, whether generated by the body like enkephalin and beta-endorphin or ingested like morphine or heroin, work at a pretty basic, reptilian level: as rewards for, or distractions from, a stressful life's various hurts. Binding to receptors found not just in the brain but also in the brain stem, central nervous system, and gut, they trigger euphoric sensations, allay pain, and produce a warm sleepiness—but it doesn't *mean* much, at least for those not predisposed to seek meaning. For every De Quincey or Coleridge who ranges the fields of paradise on stilts of opium, there are thousands of others whose experience of the drug is essentially pleasure without content. This explains in part the communal rituals of heroin—the shared works, the habitual shooting galleries—because without them there is no *story*, only a rush of bliss and a retreat of care. Eventually, addiction makes even this a negative process; the body ceases producing its own opiates: cold turkey is what we would all feel like, all the time, if endorphins did not deaden much of our nervous system. As users and their families quickly find, repeated seeking after empty relief drains meaning from the rest of life. The cliché truth of junkiedom isn't only that the drug matters so much; it's that nothing else does.

Drugs like peyote, psilocybin, and ibogaine, though—the so-called entheogens, used by traditional religions to get in touch with divine forces—seem to have an opposite effect: they create and amplify the appearance of meaning. The cliché here is the portentous revelation of the acid trip: "Everything's *related*, dude." While it's unlikely that the totem spirits actually come down to guide the shaman, or that all beings are fringed with a celestial radiance, or that the far hills are dotted with tiny hobbit-houses, nevertheless, these mistaken perceptions appear very like the mistakes the brain makes anyway in the course of its attempts to capture significance from the passing scene. In primitive life, where humans have little enough control over the forces that shape their destinies, this extension of illusion may be useful—better a false sense of divine purpose than starvation, disease, and violence *plus* existential despair. It may also have a useful role to play in more modern troubles, as narco-adventurer Daniel Pinchbeck found in his encounter with ibogaine's Primordial Wisdom Teacher:

> Much of my trip focused on my personal faults, which were displayed in detail, and my anger at myself for being unable to correct them. The answer I got was "GET STRAIGHT! DO WHAT IT TAKES TO STRAIGHTEN OUT THE SHIT!" When I argued that it really didn't matter, the answer was: "EVERYTHING MATTERS!"[69]

Iboga, the root of a West African shrub used in the Bwiti religion of the Fang people of Gabon and Cameroon, contains a number of alkaloids that produce a variety of neurological effects, from a mild exhilaration to a two-day journey of vivid hallucinations combined with an inability to move. But in addition, it appears to have the power to break opiate and other drug addictions immediately, and in many cases, completely. Much of this effect is purely chemical (it seems to work, for instance, with rats)—but at least some of the reason for its effectiveness is the overwhelming quality of the spiritual experience.[70] People report seeing their lives laid out for them in a didactic form that allows them to ask how they reached this point and to understand the effect they have had on others. Is any of it true? That hardly matters if the experience mimics what the brain would like to do for itself. As Patrick Kroupa, legendary hacker and a heroin addict for sixteen years until his ibogaine treatment, explains: "Everything can be distilled down to exactly one word: belief . . . you need to change your definition of who and what you are; to rewrite your own mythology."[71] And once rewritten, the new myth persists—more than a year later, people who have had experiences with enthoegenic drugs continue to believe that something spiritually significant has happened to them.[72]

CRIPPLING CLARITY

It is actually an error to claim that the human brain cannot function like a computer, because there have been a few that have. Jedediah Buxton could reckon the number of square inches in a farm merely by walking over it. He could also tell you how many words were spoken by each character in *Richard III*—without being able to say what it was about. Stephen Wiltshire has drawn a thirty-foot-long aerial view of Tokyo, accurate to the number of windows, based on a single short helicopter flight. Solomon Shereshevsky, described in A. R. Luria's *The Mind of a Mnemonist*, could recall every word of anything he had heard or read, although he could not remember faces, because they "were so changeable." In all their cases, as with many such natural savants, the skill they show is dearly bought; their omnivorous senses or calculating genius allow no preference for one element of experience over another, no chance to create meaning by selective discarding of the world's oppressive bounty. Like computers, they accept all input. Oliver Sacks, in "The Dog Beneath the Skin," tells the story of a student who, after too many amphetamines, briefly enjoyed the same vivid and relentless quality of life—this time through the sense of smell.[73] He noted the "happy" smell of water, the "brave" smell of stone; he could name who was in a room before he entered it. It was a huge pleasure but at the same time "a world overwhelming in immediacy," where thought became difficult and unreal.

Unencumbered by miraculous talents, unstimulated by amphetamines, our brains take as their principal role protecting us from life, by selecting and arranging fragments of experience into a beautiful but stylized mosaic. Why? Because, as probability experts point out but no layman believes, almost all of life is random. If, forced by some higher power, you had to give a true account of how you came to be where and who you are, it would read like a chapter of accidents, in which the real skill was simply saying "yes" to the right invitations: showing up for the better chances and being elsewhere for the worse. Hard work, or being a good companion, or occasional flashes of boldness—these abilities may offer a promise of control of things, but never the assurance. It could all have turned out entirely otherwise; or why else do adult people, alone in the kitchen after midnight, sit and Google old classmates and old lovers?

CHAPTER IV

Off the Rails

Human error is usually thought of as a fault in our control of technology—unlike people, machines do exactly what we tell them to do, which makes them rather dangerous companions. We are all familiar with the sense of discomfort, unfamiliarity, and mounting panic when we are in charge of a system we can't quite control. We also know the bliss of being "at one" with a machine, whether car, airplane, or musical instrument—but this bliss brings its own species of error. So how do we distinguish between the mental errors that produce disaster and the ones from which we can recover? Through probabilistic thinking: the people who are best to have in charge of machines are the ones who do not themselves think in a machine-like way.

On curves ahead
Remember, sonny
That rabbit's foot
Didn't save
The bunny

—BURMA-SHAVE ADVERTISEMENT, 1950

IN chapter III we explored the pitfalls of perception, but perception is only the beginning of the game: we have not just to see, but act. Life calls us out of the private fastnesses of our heads into a world of *things*, things we valiantly attempt to steer, start, or stop—and this opens a whole new playground for error. The English humorist Paul Jennings once proposed an alternative to the existentialist *nausée* that was then washing across from Paris: resistentialism. This "somber, post-atomic philosophy of pagan, despairing nobility" held that "*things* are against us." And it seems they still are: rather than the masters of the physical world, we remain mere "no-things," doomed to be pushed around by scornful and uncooperative objects—pianos that mock our sausage fingers; computers that develop transient but alarming hypochondria; keys, socks, and teaspoons that scurry off to their secret covens and never return.[1] There are certainly days when resistentialism seems the only explanation. Even the clearest view of the world does not necessarily give one the ability to control it; nor are the people who *do* run things exempt from potentially dangerous illusions about themselves and their expertise.

We poor naked forked creatures have invented a world of aids to make good our own deficiencies: clothes; wheels; lenses to correct poor vision; wings to release us from the shackles of gravity. Our technologies and institutions exist only to fit our needs. Why, then, should things seem against us? Perhaps because they do not share our illusions—the cognitive quirks that mark the same inventive, problem-solving minds that created them.

THE ILLUSIONS OF NORMALITY

Let's start with a test. If someone asks you whether Naomi—artistically gifted at school, radical in her politics, and with poor math SAT scores—is more likely to work as a bank teller or as a sculptor, you, like me, will probably plump for the latter and be wrong. "Likely" is a matter of probability, and statistics tell us there are bound to be far more bank tellers than professional sculptors, even among a sample that matches Naomi's characteristics. If you are asked whether she is more likely to be a bank worker or a nonprofit *co-op* bank worker, you will make the same choice—the latter—but a different mistake: co-op bank workers are a subset of bank workers. These sorts of errors are *cognitive illusions*: natural, universal failures of our frames of reference—as common and inevitable as the optical tricks we fell for in the last chapter.

Cognitive illusions are the points where everyone's apprehension breaks down and logic goes awry. It is becoming clear that these illusions are not simply the result of sloppy or haphazard reasoning. We are wrong, but we are wrong consistently, and much of this consistent error reflects how different our rich, busy world is from the pure white cubical halls of logic. Our mental resources are used elsewhere—in classification, distinction, speedy commitment—not in the induction from limited data of universal facts.

Probability, the formal science of experience, depends on repetition; it can tell us what outcomes are likely over time, or in the aggregate, but it has no way of describing the likelihood of a *single* event—the here and now. But the single event is what really interests us: we want to know, not some abstract *N*, but Naomi herself.

Logic is also absolute, while our minds seem more attuned to relative relationships; we like to reset our expectations to fit the context. It's not just that bar talk is different from church talk; arithmetic itself can depend on the frame of reference we began with. When, for instance, people try to multiply in their heads the numbers ascending from 2 to 8, they come up with an average total a quarter the size of that when they try to multiply the numbers *de*scending from 8 to 2 (and an eightieth the size of the real answer).[2] Thus small beginnings tend to small ends, but—as every government knows—you can do what you like when you start in the billions.

This essential relativity isn't confined to numbers; it also governs degrees of belief. Our confidence in what we say we know has almost an inverse relation to how well we actually *do* know it: people, even after they are briefed on exactly what such odds represent, will happily bet at 100-to-1 that more Americans die from homicide than suicide or that the potato originated in Ireland. We line up our imagined world more neatly than reality, making Italy a straight north-south leg, France a perfect hexagon, and New York level with similarly "topside" cities like London and Berlin, when in fact it shares its latitude with Rome and Barcelona.[3] Geographers would certainly not consider these errors trivial. And if you ask anyone about a topic he or she knows really *well*—heart surgery, property finance, the earlier albums of Black Sabbath—the response will show much less cocksureness. It will be nuanced and limited, with clear boundaries drawn between the likely, the debatable, and the unknown. But the same person quizzed on an unfamiliar topic will, like the rest of us, happily retail opinions that have as little relation to fact as the tale of George Washington's cherry tree.

Why? Because, no matter how ignorant we may be of a subject, we refuse to be speechless. We hate the necessity of saying, however truthfully, "I dunno." We must have a ready opinion, and pretty much any opinion will suffice: it takes an effort of will and, usually, a rigorous education to distinguish real knowledge from plausible guff. The concept of *testing against the null hypothesis*—admitting and gauging the possibility that there is only a random connection between two observed facts—may be the foundation stone of the scientific method, but it is missing from our personal reasoning, because we find the truly random almost impossible to imagine.

This makes us naturally optimistic: the same instinct that allows us to believe all our children are above average and that we are more likely than others to pay off our mortgages or avoid drinking problems and heart attacks (an instinct located, according to recent research, in the rostral anterior cingulate cortex) also influences the way we connect evidence and belief.[4] In general, we put undue emphasis on the link between *positive* evidence and our hypotheses than we should: we latch onto the visible fact to support our theory, rather than scanning the background randomness, where nothing in particular happens most of the time. The self-mastering genius of a Sherlock Holmes might notice the curious incident of the dog in the nighttime; the rest of us are only roused by a bark.[5]

False trust in the positive is no mere anthropological fancy; it can endanger lives and happiness. Imagine you are a doctor, testing a patient for a particular cancer. You know that 0.8 percent of women her age have the cancer; if she does, the chance is 90 percent that she will test positive.[6] The test also has a 7 percent chance of giving a false positive result. In her case, the test *is* positive. Does she have the cancer? The great majority of doctors at a German teaching hospital confidently thought so when the problem was put to them; several assumed the chance was 90 percent of a correct if unwelcome diagnosis. After all, the test is accurate and positive; what more could you want?

In fact, the chance the patient has cancer is just over 9 percent, which you will see if you reframe the problem in these terms: "Think of a thousand women: eight will have cancer, and the test will detect it in seven of them. On the other hand, the test will *falsely* detect cancer in seventy women. Your patient is therefore one of seventy-seven women with positive results, seven of whom really have cancer." Our clumsiness with percentages compounds our natural prejudice in favor of the scenario where evidence confirms assumptions—positive results for a positive reality.

Here's another example—it is the basic structure of any scientific experiment or practical test, stripped down to its essential formal question. If you find this easy, the world has many important jobs for you. If you fail, then welcome to the great majority.

You have four cards in front of you, marked with letters on one side and numbers on the other:

The hypothesis, or rule, you want to test is this: "If there is a D on one side of the card, then there is a 3 on the other." *Precisely* which cards do you need to turn over to determine if the rule is true?[7]

"D and 3" is the near-universal wrong answer. In fact, if you want to be certain of eliminating not just the false positive result ("D, yet not a 3") but also the false *negative* ("not a 3, yet D"), you need to turn over D and 7. The desire to remain in positive territory is so strong that, of the many students who have taken this test, the few who managed a correct answer were usually only those who had taken a class in formal logic and laboriously applied its *rule of contrapositives* ("if A, then B" implies "if not B, then not A"). Few of us carry that handy crutch; we have trouble enough simply keeping multiple negatives in our single mind—as we show when we say things like "I could care less" or "nothing is too trivial to ignore."

COMPLEX SYSTEMS, SIMPLE MISTAKES

All right, then. We humans are overconfident in our answers, blur the distinction between in-depth and hearsay knowledge, and concentrate unduly on positive confirmation of our hypotheses. But all this comes out of academic trick questions; in the real world, we have the rich context that is missing in these sparse and sneaky hypotheticals. How could these cognitive illusions really matter?

A jumbo jet can weigh 875,000 pounds at takeoff, 400,000 pounds of which is fuel packing an energy equivalent to 1.7 kilotons of TNT, or one seventh of a Hiroshima bomb. The aircraft has six million parts, 170 miles of cabling, seats for 524 passengers . . . all, you reflect morosely as you crunch your pretzel-style snack, at the mercy of two guys up front rehashing the Lakers game. Yes, they *know* their jobs—they have been selected and trained with a care few earthbound roles demand—but what kind of knowledge is this? And (a surreptitious tug on the seat belt) what if it fails?*

British Midland Flight 092 from London Heathrow to Belfast on

* You will notice a lot of examples from aviation in this chapter. This is not to say that flying is inherently more prone to error than other technical realms—the reverse is probably true. It is, however, an area where any mistake is so costly that each is well documented and carefully studied.

January 8, 1989, was a routine shuttle route, operated by an established, safety-conscious airline using a brand-new plane, a Boeing 737-400.[8] Even after a turbine blade ruptured in the left engine during the final phase of the climb, starting an engine fire and sending severe vibrations and a smell of burning through the aircraft, the situation was far from critical: a 737 can fly well enough on one engine to seek out and land safely at an alternative airport. The only essential thing to do was to cut fuel to the ailing engine and snuff out the fire.

The pilot asked the first officer which engine had the problem. "It's the le . . . it's the right one" was the fateful reply. On that basis, they then throttled back the *un*damaged right engine (noticing, coincidentally, a reduction in the smell of burning) and headed for East Midlands airport using only the damaged engine. When this finally gave out three miles from the runway, they were flying too slowly to restart the right engine and they crashed in the median strip of the M1 highway, killing 47 of the 118 passengers.

What made the first officer answer as he did, when the instruments would have shown fluctuating power in the left engine and smooth running in the right—and when anyone in the cabin could have told them there were flames in the left exhaust? Not panic, not ignorance, but *knowledge*: he knew that the air-conditioning systems for all earlier 737 models had their intake in the right-engine compressor—so a smell of burning in the cabin air would necessarily mean fire on the right. The new model they were flying, however, drew its air equally from both sides; but the first officer did not know that. He took the positive evidence of his nose to confirm a positive, wrong hypothesis.

So many of the errors we make in this complex, risky world follow the same pattern: increased familiarity with a task breeds positive hypotheses, assumptions not just about how things run normally but also about how they "normally" go wrong. In trying times, we seek and usually find confirming evidence to support one of these hypotheses—we assume the predicted—rather than asking whether an unknown cause may be responsible for the elements *missing* from the familiar picture.

The better we get at a job and the more polished our handling of routine, the less likely we are to notice baffling anomalies: the more you can do with your eyes closed, the bigger your blind spot when they are open.

Thus the captain of the *Royal Majesty* cruise ship, knowing that it was regally equipped with GPS and an automatic depth alarm, *assumed* that it was protected against failures of navigation. He had no need to worry about the notorious Nantucket Shoals—until when, the ship now firmly stuck on them, he discovered that the GPS antenna wires had come loose and the depth alarm was mistakenly set to zero. The control-room staff at the Three Mile Island nuclear plant *assumed* that rising coolant levels in the pressurizer above the reactor core meant too much coolant around the core—until the incoming shift finally noticed that rocketing temperatures and overflowing sumps meant steam, not water, was pumping through the system and the core was exposed and melting.[9]

Rarely is such a disaster the result only of individual failure—a job anyone else might have done better. Human error, like human achievement, usually combines elements of the personal and the social. Group enterprises can develop collective cognitive illusions just as dangerous as an individual's. The rough-and-ready world of 1980s oil exploration—where you didn't read the damn manual, you asked the guy who knew—largely contributed to the capsizing of the theoretically unsinkable semisubmersible drilling rig *Ocean Ranger* during a storm off Newfoundland. When a freak wave shorted out the ballast control panel, there was no "guy who knew" about such an unexpected situation. The crew wrangled the valves manually and haphazardly, trying out hearsay fixes and half-remembered work-arounds; had they done *nothing*, they would have had a better chance of survival.

No one, by contrast, could call the development process for a frontline air superiority fighter "rough-and-ready," but the industry's culture of widespread delegation of responsibility can open loopholes for flawed assumptions to slip through. On February 11, 2007, a flight of six F-22 Raptors left Hickham Air Force Base in Hawaii on their way to Okinawa.[10] Partway through the journey, and at almost exactly the same moment, every plane completely lost its navigation, fuel-management, and communication systems—leaving them, despite their $125 million price tag, as dumb and blind as crop dusters. Luckily the weather was clear and the flight was able to limp home, meekly tagging after its accompanying tankers. What secret enemy could strike such a devastating blow? The international date line. Nobody at any of the more than one thousand companies that worked on the Raptor program had remembered that in traveling west across the Pacific, the date shifts forward—an apparent leap through time that caused the onboard software to go into a terminal fugue. Sometimes, as Ogden Nash pointed out, "too clever is dumb."

THINKING THROUGH YOUR HAT

We are, remember, relativists: context exerts a subtle but irresistible pull on our judgment, a pull so strong that it can draw us right out of the world of physical consequences. On the evening of January 27, 1986, a telephone meeting brought two incompatible frames of reference into the same room.[11] The first belonged to a group of engineers, appalled at the prospect of subjecting a potentially flawed design to unprecedented stresses after its testing had yielded inconclusive evidence. The other belonged to a group of project managers, equally appalled at the prospect of jeopardizing a vast program's reputation for dependable, routine delivery in front of its political paymasters and the world's public. The engineers worked at Morton Thiokol, the rocket-engine company; so did the managers. Waiting on the other end of the phone during this five-minute time-out from a conference call were their customers at NASA. The launch of the space shuttle *Challenger* was due the next day; the suspect system was the now infamous O-ring seal in the solid rocket booster case joints.

At the pivot between the increasingly angry engineers, who wanted to postpone the flight, and the frustrated managers, who felt it had already been postponed too often, sat Bob Lund, Morton Thiokol's VP of engineering. He had received a memo from the O-ring team that spelled out exactly what was at stake: "catastrophe of the highest order—loss of human life." He himself, having seen the engineering data, had recommended against launch on the following day. But now he had just been listening to Larry Mulloy at NASA, who found the evidence unconvincing and the potential for delay unacceptable: "My God, Thiokol, when do you want me to launch? Next April?" Now, too, he was sitting next to his general manager, Jerry Mason, who had not been closely involved in the yearlong agonizing over O-ring erosion (a year in which, it should be pointed out, ten shuttle missions had flown successfully). Mason, turning a dead eye on the engineers, said simply, "We have to make . . . a *management* decision." The managers huddled at one end of the table.

Mason asked whether he was the only one who wanted to fly. "What do you think, Bob?" Lund hesitated. Mason added, "Take off your engineering hat and put on your management hat." Like some magic cap in a Grimm tale, the management hat subtly transformed reality, changing the landscape of risk from a place of rare but nigglingly significant dangers to a place where things usually work and decisions have to be made, because so much is at stake. Under its influence, Lund agreed to recommend launch, and NASA snapped up the recommendation almost without question. The next day, the O-ring failed, seven astronauts died, and the space shuttle program lost its raison d'être.

What is the mystic power of the management hat? It makes the wearer subscribe to a different theory of probability. Engineers follow the classical mathematical model, where every trial of an unchanged system suffers the same risk. O-rings, like dice, have no memory. They will burn through some day, according to their innate fallibility—but that day is not advanced or delayed by what has gone before. Management probability looks at *results*, not causes, and so is closer to a gambler's belief in "streaks" or "hot hands." In Richard Feynman's appendix to the official report on the *Challenger* disaster, he says that NASA managers seemed to assume that the repeated use of a faulty system without failure actually *reduced* the risk in subsequent use. But, as he pointed out, "when playing Russian roulette the fact that the first shot got off safely is little comfort for the next."[12]

The gap between engineering and management probability is not really that between right and wrong, though—between principled whistle-blowers and sleazy suits. It has most to do with whether the experience is going to be repeated: after all, you're highly likely to *survive* at Russian roulette if you play only one round and get to go first. Some situations reward the risk-taker and some don't. The reason we have different hats to think with is that no single viewpoint can take in all of a complex world: the painstaking, deliberate qualities of a good engineer can wreck a large social enterprise as surely as the damn-the-torpedoes habits of a successful manager can cripple a complex technology project. This implies, therefore, that one reason two people can, in the same situation, commit contradictory errors is that they have different models of probability. Their fundamental assumptions about the world depend on their different experiences and expertise.

The engineering hat therefore creates illusions of its own, the most alluring of which is the vision of perfection. In the early 1990s, Denver International Airport saw an attempt to create the baggage handling system of the future—a $186 million wonder that tracked every suitcase through an all-knowing mainframe and shunted each one through twenty-six miles of track using automated trucks, extensible conveyors, and pneumatic switches. It was perfect—in theory.[13] In practice, the loaded trucks often couldn't manage the space-saving sharp corners and simply tipped over. Bags lay in limbo as ambiguous data-points about which every question could be answered except the simplest and most important one: *where is it*? Design perfection had left no room for error. The airport struggled with the system for ten years and finally went back to using people. "It wasn't the technology per se," comments Richard de Neufville, an engineering systems professor at MIT. "It was the misplaced faith in it. The main culprit was hubris."

BECOMING EXPERT: ORGANIZE, SIMPLIFY, THEN FLOW . . .

What becomes clear when we examine error is that it generally stems from our attempts at expertise. We don't just flounder around randomly; our mistakes are determined by how we try to filter randomness *out* of experience: through our various assumptions, heuristics, and models. This means that to really mess something up we must have gone through many of the same mental processes as are involved in becoming very good at it.

Driving is the *other* thing that no man will admit he does badly; and learning how to do it can be, for many, a similarly baffling and anxious experience. The quadrupedal ritual of key, mirror, indicator, clutch in, shift, clutch out/hand brake off/accelerate (have you left the indicator on?) initially appears as difficult to reproduce as a Buddy Rich drum solo. Yet most people advance past this stage and essentially *forget* the individual actions in the sequence—and the more expert they become, the larger the units of experience they stash in the unconscious. When you ride with Lamborghini's test driver, you'll find his focus far beyond his own swift, sure movements; instead, it fixes on upcoming curves and cambers, tire adhesion, torque versus speed profiles, oppressive safety legislation, and favorite restaurants—as the landscape blurs by and your own feet dance on phantom brake-pedals.[14] His world—its timing, its variables, its dynamics—is not the same as yours.

Why does expertise change one's physical reality? Once more, it's a matter of the brain's fundamental need for economy in using its resources. Fourteen-month-old children, at that age when learning about the world becomes a torrent of new acquisitions, still have only three available places in their working memory; to handle any more objects than that requires categorizing them.[15] From earliest childhood, therefore, we learn to simplify by making equivalents between things and sorting them into *taxonomies*: Teeny and Ruff are both dogs, a dog and a horse are both animals, animals and people are both alive. These categories, unlike mathematical or logical sets, tend to be weighted in our minds toward more likely or more representative examples: cows are *more* "animal" than turtles; pigeons are more "bird" than penguins; oneself more "alive" than, say, lichen. The purpose of this folk taxonomy is not to build up an accurate picture of the world's dense variety, but to slash a shortcut through it, to ease planning and decision by letting one thing stand for many in the mind.[16] This phenomenon goes a long way toward explaining why we are most sure of what we know least: our one mental example can cover a whole phylum of phenomena. Significantly, where such simplification would hinder us from dealing successfully with the world, we avoid it: shepherds can tell individual sheep apart, and a good batter will know a slider from a cutter as it leaves the pitcher's hand.

As we extend our taxonomies from home territory into the unknown (so that, when first you see a wildebeest, you will still know it is an animal, and more cow than cat), we are also forming theories about how the elements of these categories are likely to behave: a grammar to fit nouns both familiar and unfamiliar. The process is known as *inductive inference*. Unlike deductive logic, it doesn't derive the qualities of an individual example (Socrates) from general rules (mortals), but instead predicts whether the qualities of an example will apply to the whole class of which it is a representative ("Socrates loves to talk—do *all* mortals love to talk?").

Outside the Castle Adamant of mathematics, such inductive predictions can never be proved; they can only be probed through repeated, nested judgments of likelihood. This is childhood's primary business, and a lot of the apparently silly questions we like to ask small children ("Do you like cookies? Does your hamster like cookies? Does your hamster like *you*?") are actually teaching aids for coming to inductive conclusions about how stuff usually sorts and works. Interestingly, a recent (though small and none too rigorous) study at the University of Leicester suggests that people who have suffered great upheavals in childhood (illness, poverty, persistent bullying, death of a family member) are more gullible as adults; the implication is that overwhelming misfortune gave them less chance to develop a personal sense of probability—a world that would take away your mother when you are only three seems essentially, cruelly, arbitrary.[17]

At the level of individual observations and decisions, the inductive process can be modeled through *Bayesian calculation*, a branch of probability theory that describes how a single event can change our prior assumptions about a class of phenomena. Unlike classical, dice-rolling probability, Bayesian calculation does not require a large number of examples to come to a strong provisional conclusion, so even where we have little experience, we can at least have an opinion. In other words, Bayesian probability wears the management hat. This is probably useful for an animal with limited time to make sense of a complicated world, and it may also explain why becoming a scientist is so much more difficult than becoming a manager.

Bayesian probability is what lets us take elements out of play and concentrate our limited mental resources on the bigger game, replacing conscious stepwise sequences of observation and decision with *heuristics*—recipes, maxims, and rules of thumb that get the job done quickly and efficiently. Thus, mathematicians often marvel at how someone running for a fly ball is actually solving simultaneous differential equations in four dimensions—a task that would take days, if feasible at all, using pencil and paper. Are left fielders geniuses, then? Well, maybe not: careful study has shown that even experienced players cannot predict where a ball will land unless they are running toward it—and even then they tend not to run in a straight line nor at a constant speed.[18] If, though, you model what they're doing in terms of the heuristic "look at the ball and run, adjusting direction and speed so that the *angle of gaze* remains constant," then outfield choreography suddenly becomes comprehensible and we can forget about signing up a lot of ballplayers for the math department.

The more familiar you are with the quirks and variables of something—that is, the more it *interests* you—the more sophisticated your heuristics. "Draw your seagulls like upside-down mustaches" will not do for a Leonardo; Murphy's Law might have been good enough for Captain Murphy, but his colleagues on the MX981 rocket-sled project felt he should have taken a more nuanced view. One survivor insists that the original law was better formulated by the project leader, Dr. John Stapp, as a cheerful reminder of probability's *ergodic principle*—"whatever *can* happen, *will* happen, eventually"—rather than a gloomy reflection on the cussedness of things.[19] Marxist historical theory is a usefully simple heuristic if your purpose is to start a dorm-room argument, gain tenure at a minor college, or establish a peasant insurgency. But if historical change actually interests you, your rules, while still encompassing whole centuries and nations, will try to include some of the variety and fascination of the human spirit: the sense of innumerable biographies within the larger sweeps and surges, the repeated questions that have no answer. It's not an easy task; as the historian Peter Gay said, "Many historians have heard the music of the past and then transcribed it for the penny whistle."[20]

His point applies to any error of cognition: since we *must* simplify experience to do anything with it, the quality of that simplification will have a critical bearing on whether life will be a stately dance or a series of pratfalls. Becoming good at something demands a higher-order encoding of experience that assigns our limited mental resources to the points where they can exert the greatest leverage.[21] Once that encoding is made, it is permanent—as easily accessed and unexamined a skill for the adept as shoe-tying or whistling for the rest of us.

Thus the expert can often sound as arrogantly simplistic as the novice; the only difference is that his arrogance is justified. When the chess grandmaster José Raúl Capablanca said, "I see only one move ahead, but it is always the correct one," he meant that the shapes of potential games, the millions of permutations that computer chess programs must grind through before making a decision, were so ordered in his rigorously trained mind that he could guess without effort the outermost likely implications of each next move—he could see the tree in the acorn. One opponent, subsequent world champion Mikhail Botvinnik, said, "Capablanca didn't make separate moves—he was creating a chess picture."[22]

The fourteen-year-old Mozart, too, absorbed his music very differently from the rest of us, but also differently from the phonographic memories of savants and prodigies. When, after one hearing, he was able to transcribe Allegri's nine-part choral *Miserere* (the score of which the Vatican had kept secret for fear lest other, undeserving audiences might enjoy its heavenly harmonies), he was not performing a feat of recall but of understanding: the shape and intent of the piece were clear to him, allowing him to sketch in the details accurately from his own resources. We say "resources" because skill with heuristics is not some miraculous gift—it is built up gradually through intense absorption in the work. The only *inborn* talents that mark the future genius are the ability to concentrate and the burning desire to master; all the rest is acquired the same way everyone else learns new skills: through directed, repeated experience. Don't forget that the fourteen-year-old Mozart already had more than ten years' practice at the highest level of European musical culture.

True knowledge also includes knowing what you *don't* know. "El Maestro," Juan Manuel Fangio, became one of the most successful racing drivers in history because of his near-perfect knowledge of limits. If, on a given curve, his car was likely to come unstuck at 95 miles per hour, he would take it at 93, while lesser talents might settle for 90. He never pushed through the envelope, but he explored it completely. He knew every pertinent quality of his machine, the track, and his opponents, setting a standard of cool professionalism that still marks the sport. What he did *not* know, however, was the temper, skill, or sobriety of civilian motorists, so he insisted that all long-distance drives with his family should start at three in the morning, because then the roads would be emptiest. As known quantities, cars gave him the chance to excel; as unknowns, they made him nervous.

Old teachers grow tetchy and disappointed, not from the effort of teaching, but because so much that happens in a classroom isn't about teaching at all: "Is this going to be on the test?" "Mrs. Adams never made us learn this" "Why are you *asking* me things? Just tell me what I'm supposed to know!" Real learning, real expertise, occurs in an entirely different mode: the pleasurable expectation of being challenged, tested, and surprised. The painter and printmaker Hokusai, having begun sketching at the age of six, calmly explained in later life:

> Nothing I did before the age of seventy was worthy of attention. At seventy-three, I began to grasp the structures of birds and beasts, insects and fish, and of the way plants grow. If I go on trying, I will surely understand them still better by the time I am eighty-six.[23]

Hokusai died at eighty-nine, protesting: "If I had another five years, even, I could have become a real painter." Beethoven's late quartets, Rembrandt's late self-portraits, Milton's late lyrics—well into old age, the active mind rejoices in forays beyond the known and comfortable, setting its eyes on the exceptions to well-known rules. The master does not ask "what am I *supposed* to know," because he already knows that all such supposition is false: even his firmest assumptions, like all probability judgments, must suffer occasional shaking. And what is true of the mental world is equally true of the physical. Tomorrow Tiger Woods might stop driving the ball as well as he can when playing at his bewildering best; but, as a master, he would find this problem *interesting*, not just horrifying, and experiment with ways to compensate—by, say, taking only iron shots off the tee, sacrificing length for accuracy, as he did to such devastating effect at the 2006

Open Championship. His heuristics have not ossified into laws: they have the toughness to absorb the rare and unexpected.

How can we be sure that the mark of expertise is the combination of more complex taxonomies and more sophisticated heuristics? At the most basic level, we know it because expertise leaves traces on our brains, as revealing to the informed investigator as the brown mark on a violinist's neck or the asymmetrical buttocks of a champion fencer. Each neuron's synapses operate under a harsh and simple rule: communicate or die. Babies born with cataracts will never be able to see through the affected eye unless operated on almost immediately; later corrective surgery will leave a functional eye but a brain indifferent to that eye's output.[24] On the other hand, growth in capabilities can rearrange and extend brain connections, in the phenomenon known as *plasticity*. It has been shown at work through a rather spooky experiment from the Howard Hughes Institute, in which transgenic mice with fluorescent brains and little windows set into their skulls produced glowing new connections in their sensory centers as a result of having to explore the world with trimmed whiskers.[25] The same happens with humans who acquire complex skills or detailed knowledge, even in adulthood: string players show a greater development of the regions of the motor cortex that control the left hand. London taxi drivers, who cannot get a hackney license without demonstrating mastery of "the Knowledge"—a memorized account of every street and route in the metropolis—show expanded areas of the hippocampus governing navigation and spatial memory.[26] The more detail, the more connections . . . and the more we tend to see the world in terms of our own specialty. As Jack Loizeaux, founder of Controlled Demolition Inc., the world's masters of building implosion, once said: "I know it sounds terrible, but even when I'm on vacation and I see some beautiful place like the Taj Mahal, automatically I start figuring how I'd set the charges to drop it."

Automatically is the significant word. It's not just implosion, driving, or golf; expertise moves the heart of *any* activity into that effortless, non-introspective zone of immersion that the psychologist Mihaly Csikszentmihalyi calls "flow experience." Time slows and things lose their opposition to the self—it is as if they were cooperating, like the target in Zen archery. "I truly loved those fastballs," said the old batting star Goose Goslin. "*Zip* they'd come in, and *whack*—right back out they'd go."[27] This feeling of harmonious action is perhaps the greatest pleasure we can know, because it erases the essential loneliness of the individual mind. It is what gives that golden tinge to childhood days in the sandbox and what explains why we so willingly do as hobbies—fishing, quilt making, driving steam engines—what our ancestors would more accurately call work.

The opposite feeling, though, is one of the worst we can imagine: the athlete's *choke*, the actor's *dry*, the writer's *block*. Here all is isolation, the inexplicable recalcitrance of things, time slipped agonizingly out of joint. Every movement becomes conscious, deliberate, leaden. To see another, even an opponent, in the grip of the choke is never pleasant: we know the horror too well from own nightmares. Because expertise and practice have made so much of the business automatic, there is no ready, rational explanation for why the old fluidity has suddenly frozen—and no logical method to rebuild technique from first principles. Choking is the one big reason why acting and sports remain such haunts of superstition and personal magic.

Naturally, you don't want a choker in charge of your critical enterprise; but then, you may not want someone in the raptures of flow experience, either. "Bud was very at ease in the airplane: a situational awareness type of guy—among the most knowledgeable guys I've flown with in the B-52."[28] In June 1994, Lieutenant Colonel Arthur "Bud" Holland was chief of the Ninety-second Bombardment Wing Standardization and Evaluation Section at Fairchild Air Force Base in Washington. He could do things with the hundred-ton Stratofortress that made it look like a fighter plane: steep banks, near-vertical climbs, wingovers, and thunderous low passes, the great ash-gray underside only feet from the ground. Although far outside allowable procedure, these aerial stunts made him a favorite at air shows, photo opportunities, and change-of-command jamborees. Though perhaps not a favorite with everyone: several junior officers, after one bent rule too many, had refused to fly with Holland. His response was that of a man who knew the gods were on his side: he called the doubters "pussies" and laughed "a good belly laugh." It was widely believed on base that Bud Holland planned to be the first man ever to roll the B-52.

How, as many people asked at the time, did he get away with such indiscipline? "I don't know," was the usual answer. "He just . . . does." Yes, there were some symptoms of institutional malaise—the unit was losing its bomber status, commanders came and went too regularly—but perhaps there was also a grudging sense that Bud Holland represented *what flying is all about*, the belly laugh in the face of danger that made him kin with the pioneers, when all between you and the sky was canvas and string and the only neck you risked was your own. In a checklist-and-clipboard world, his stick-and-rudder talents somehow gave him special license.

On June 24, 1994, Holland was piloting *Czar* 52 in a practice run for a forthcoming air show. He had briefed for a routine involving extreme angles of bank and pitch and been told these were unacceptable—but, as always, he was doing them anyway. His copilot was a colleague who had repeatedly tried to have Holland grounded; in the back was a safety observer celebrating his official last flight—his wife was waiting with champagne on the flight line. As they prepared to land after the practice, they were told to do a "go-around" because of another aircraft on the runway.

The video has a nightmare's fascination: the giant object, as impossibly airborne as a flying locomotive, seemingly hangs in the air, banking through 360 degrees with its left wingtip almost lower than the control tower. Steeper and steeper the angle of bank grows, until . . . it passes the vertical and all lift disappears and *Czar* 52 simply slips from the sky, crumpling almost demurely into a carpet of flaming wreckage. It's a fair bet now that no one will ever roll a B-52.

So expert pilots can cause disasters, just like inexperienced ones; knowledge can kill people as surely as ignorance; the culture of lab coats and calculator holsters can get things as wrong as that of hairy chests and hi-vis tabards. What can we do about it, other than stay home and creep under the covers? Can we design life to save us from ourselves? Not very easily, because the *offset effect* stands in the way. Our pursuit of flow makes us boost our risk-taking in response to each new safety feature. When antilock brakes were first introduced, cars fitted with them actually had *more* crashes—they conveyed a false sense of invulnerability. Bike riders wearing helmets attempt more dangerous mountain trails, more infants are left unattended in bath seats designed to prevent drowning, and, frustrated by the hard-to-open childproof caps on drug bottles, more of us simply leave them off.[29] Nothing is either safe or hazardous, but our using makes it so. In the words of Pogo: "We have met the enemy, and he is us."

HOW TO BUNGLE BETTER

If you look up "human error" in the professional literature, you will find that it has been largely replaced with the term "human reliability." This is not mere political correctness; it stems from the realization that mistakes are an irreducible subset of performance. On this earth, humankind is only perfectible up to a point; designing your systems on the assumption of uniform excellence is to offer up your whole enterprise as a hostage to fallibility. No one is necessarily *at fault* when mistakes are inevitable: "In tragic life, God wot, no villain need be!" Error is not always sin; nor is a slip the same as a blunder, a lapse equivalent to a muddle. So rather than simply warn, condemn, and forbid, human-reliability assessment seeks to break down high-risk activities into qualitatively separate phases, each with its characteristic potential errors. It then suggests modifications to the activity, not just to reduce the number of errors, but also to lower the magnitude of their impact at each phase. James Reason, doyen of the field, calls this the "Swiss cheese model": if you can keep the holes in several slices from lining up, then the overwhelming disaster may not slip through.[30]

What, then, *are* our characteristic mistakes—the sort that multinational companies are so sure we will commit that they pay expensive consultants to come and rewrite their processes? The most prevalent and basic are errors of *perception*: we simply miss the evidence that things are going wrong. This is as true in most aircraft accidents as it is on the drive home from the office: "I didn't see it." "He came out of nowhere." *Timing* is another major source of difficulties; just as we all too easily link events that happen sequentially into cause and effect, we fail to recognize genuine links if there is too much lag. When your Internet connection is running slow, you tend to hit the buttons repeatedly, inadvertently buying those tickets to Orlando four times over, because you saw no immediate response to your action.

In aviation, this impatience with delay can cause pilots to fly perfectly sound planes straight into the ground: every aircraft has a characteristic undulation, a slow cycle of rise and fall as it travels along, called its *phugoid motion*. When the pilot is unaware of this, or when the plane's control-sensitivity is not tuned to the cycle, the response from the cockpit to phugoid motion can actually amplify it: stick forward doesn't kick in until the next nose-down; the instinctive pull back exacerbates the succeeding climb. In *pilot-induced oscillation*, the plane begins to porpoise through the air—where, sadly, it often fails to remain.[31]

Nothing is likely to break us of our poor response to unexpected timing, since it draws on some very ancient habits of the vertebrate brain. B. F. Skinner's work with pigeons began during World War II, when he trained them to work as the guidance system for smart bombs by feeding them each time they pecked at the image of a Japanese battleship. Peck, then feed: cause and effect. After the war, Skinner extended this idea of conditioned response into the broad, chilly doctrine of behaviorism. But in one experiment he reversed the principle by giving the pigeons the usual bar to peck yet feeding them at the same interval *no matter what they did.* The results were remarkable: the pigeons developed superstitions, long and complex dance routines that, they apparently believed, "made" the food come out—because it had seemed to work last time.[32] By disconnecting timing from reward Skinner had effectively removed common sense from the situation—and if you think humankind is above the pigeons in this, just watch any big-league batter's habitual motions when he comes to the plate.

Life is seldom as wantonly cruel to us as Skinner was to his pigeons, so we don't always replace common sense with magic. Nevertheless, we retain a full arsenal of other techniques for messing up. James Reason suggests a hierarchy of blunders, from simple physical *slips*, where you fail to apply your skills properly; through *rule-based mistakes*, where you apply the "right" response to the wrong situation; to *knowledge-based mistakes*, where you misunderstand the whole situation and decide to do something you really shouldn't. He points out that the further up this cognitive chain we go, the harder it becomes to recognize and recover from an error. Why? Because of our tendency to default to the banal: to assume that a given situation conforms to other situations that we believe are similar or "usually" happen. A slip, such as hitting the accelerator instead of the brake, is in itself a familiar situation: we recognize and catch it quickly. A knowledge-based error, like mistaking half an eight-lane freeway for a four-lane boulevard and heading down it the wrong way, involves a whole structure of wrong assumptions about normality: we are likely to suppose "these other drivers are crazy" long before the dreadful truth sinks in. We have gambled on this being a banal situation—and lost.

Reason's hierarchy also explains why experts, especially those with strong self-belief, can still be prone to error. Believing that "I can handle it" (or yet more dangerously, "*only* I can handle it") is a personal banality: the assumption that overcoming the unexpected is normal for me. What this means, of course, is that the unexpected itself stops serving as a trigger for caution or self-questioning—as the crew of *Czar 52* found out.

Knowledge-based mistakes, however, extend beyond poor self-knowledge or overconfidence. Studies in the oil and gas industry suggest that for a major accident to develop, several distinct unsafe acts have to occur in sequence: in Reason's terms, the Swiss cheese has to line up just so. The effect of this is to isolate, in the mind, any one of those acts from its potential for disaster: you could drink thousands of Cokes in the control room before you inadvertently spill one into the main computer; you could use your cell phone at the tank farm for years before some undetected fuel leak makes this an explosively fatal mistake. Given the brain's drive for efficiency, your personal heuristics are very unlikely to take account of such remote risks. Moreover, it is difficult for any one person to understand how probability works over a whole enterprise: a working lifetime lasts around eighty thousand hours, so an accident that occurs every million or so man-hours is likely to appear only once every thirteen lifetimes—long enough to be filed under "don't you worry about that, sonny." But Royal Dutch Shell has over a hundred thousand employees, so seen from headquarters, that "once every million man-hours" comes pretty close to "once a day." The challenge for corporate safety planners is to preserve the collective from events that are effectively meaningless for the individual; this means breaking down personal intuition and replacing it with prescriptive routines—thus making the health-and-safety people even *less* popular on site than they were before.

IT MUST BE SO, BECAUSE I WANT IT TO BE SO

Beyond knowledge-based mistakes lies a forest of error yet more dark and formidable, because it is closer to the unexamined center of our mind, where desire breeds monsters out of logic: the realm of *motivated reasoning*.

In late 1916, the French general Robert Nivelle was a dashing, unconventional artilleryman with an air of invincible energy. His chief of staff, Colonel d'Alenson, was in the final stages of tuberculosis and filled with a feverish desire to do something great for his country: "Victory must be won before I die and I have but a short time to live." Both men became convinced that the key to victory on the Western Front was an attack of utmost aggression by, essentially, the entire French army, thus transforming the static slow-bleeding horrors of the previous two years into triumphant breakthrough during forty-eight hours of unprecedented violence. The force of their self-confidence relieved beleaguered politicians and swept away the counsels of caution: on December 13, Nivelle was given supreme command and began to prepare his grand offensive.

Four months, though, would pass before the jump-off—and much can happen in that time. Nivelle's conviction that his attack would be irresistible meant little care was taken to keep it secret, although it is never a wise idea to assume that a German staff officer will spend four months merely trembling in anticipation. The news that seeped the other way—that the Germans were digging into caves, straightening their defensive lines, bringing up yet more machine guns—was dismissed or ignored. Even the fact that the entire plan of attack had accidentally fallen into enemy hands twelve days before the assault was considered immaterial: five thousand artillery pieces, 1.2 million men, and 170 million rounds of ammunition would prevail against any possible defensive preparation.

The fact proved otherwise. Blundering in the rain up the steep slopes of the Chemin des Dames, the bleak limestone ridge at the center of the German position, French troops were blown back by well-prepared machine-gun fire, helpless as starlings in a gale. There were forty thousand casualties on the first day, but Nivelle insisted this was merely a test of his will: *encore de l'audace!* Before the truth sank into that handsome head three weeks later, nearly a sixth of France's army was gone. Widespread mutinies broke out; the cautious Pétain replaced Nivelle—and a reactive habit of cynical pessimism was engrained in French minds that would extend its baleful influence up to and beyond June 1940.

It would be perversely comforting to think Nivelle a fool, but he was not. His disease, motivated reasoning, is one that the most intelligent people can catch, because it undermines a deeper mental structure than intelligence: cognition itself. In a paper that reviewed most studies on the subject, the Princeton psychologist Ziva Kunda argued that motivation, the desire for a particular outcome, exists well before such complex, time-consuming tasks as observing, remembering, and planning.[33] We usually have a view *before* we have information—which explains why, when you lack conclusive evidence, a good technique for making a difficult decision is to flip a coin and then see if the result makes you wish it was best of three instead.

This state of things may be illogical, but it makes sense when we think of perception as a thing we *do*, not a passive absorption of reality. All action requires motivation—if you weren't motivated, you'd be dead. In perceptual terms, motivation is not very different from expectation, through which we divvy up our scarce sensory resources. As we saw in chapter III, simply *looking* is an active, selective process, involving choice of objects, assessment of surprise, direction of attention, and division of the world into Important and Invisible. So it's not surprising that our opinions can shape not just what we think but also what we see. Some claim that this is a useful evolutionary adaptation: given that we have so little influence over our circumstances, motivated illusions might actually be good for our mental health, because they make the world we perceive conform to what we already believe. Better that than a lifetime spent being mocked by reality. In some studies, the depressed seem to have more accurate views of their place in the world—which is hardly likely to relieve their depression.

Certainly our mistakes generally err toward the self-serving, whether it's a case of researchers "finding" that the data fits the curve or taxpayers discovering that they magically owe the government less than they'd thought. These aren't necessarily *lies*—just accurate reports from a parallel, more desirable universe, which suggests why people caught out in them are so often sincere in their protestations of innocence. But the world doesn't *know* how we want it to be, so why do we cling to self-serving conclusions in the face of conflicting evidence? Simple: once we have a conclusion we don't *see* the evidence, or we downgrade it—whether it's a smoker telling you his chimney of a grandmother lived to be ninety-nine or an oil company executive telling you that climate-change science is "flawed."

No one, obviously, chooses to be a slave of illusion—we would like our reasoning to be accurate, just as we would hope our motivation is honorable. The problem is that "accuracy" has two guises in the mind, and the first, the default version, is introspective: checking your opinions for consistency with your own experience. This is bound to be problematic, because if your prior motivation can select what you perceive, it can certainly select what you recall. Moreover, this noisy world is *full* of evidence, enough to let almost any opinion live at ease with itself. Selective perception confirms our prejudice, then selective memory reassures us that this prejudice is consistent with experience—and it all flows along automatically, leaving us with a clear conscience.

That is why we are often more accurate in judging other people—what they are like, what they ought to do—than we are about ourselves. Consider the requirements of a new job. If you were hiring someone else, you'd doubtless say each of the candidates had "some good points, some bad points—it's a matter of proportion." If you yourself are going for the interview, however, your motivation, your simplified prior opinion of yourself, will usually provide you with a biased résumé of your experience, either good or bad: "they're bound to take me—look at all my accomplishments," or "they're bound to turn me down—nothing I've done is important." It's as if we all carry with us an inner adolescent, bumptious or gloomy. This contrast between real experience and the tidy museum of recollection is striking, even for people whose deeds are a matter of historical record. Chuck Yeager, the man who broke the sound barrier, now ruefully admits, "You tell it the way you believe it and that's not necessarily the way that it happened. There's nothing more true than that."[34]

For some, though, there *is* something more true than that: the ever-present judgment of an abstract creed, a body of received texts, the shared beliefs of a community. The heavy-lidded satisfaction that marks the graduate of seminary, madrassa, or yeshiva is that of someone who has gained a standard of accuracy beyond ignorant questioning—one which also gratifyingly confirms prior assumptions. Even the most sincere come to faith as *seekers*; their reasoning, necessarily, is motivated—and their ability to select their material makes any appearance of objectivity an illusion. Their minds play only home games: since faith motivates their reasoning, their reasoning can make no sense to those outside the faith. Though all are intent on the same objective—obedience to divine law—the devout necessarily lack a common standard of accuracy: hot Hassid and cold Litvak have no shared premises; the mystic Sufi will never convince the dour Wahhabi; the Propaganda Fidei cannot judge the tenets of Pentecostalism. The great advance of the Renaissance, even before the rise of science, came when we stopped trying to convince each other through quotation and turned instead to evidence, present and real, as the basis for mutual understanding.

Fortunately, there is a second method for achieving accuracy—one that engages a completely different cognitive procedure and, because of this, works much better. Psychological studies where the subjects, having made a decision, must then *explain* their thought process to others, or must commit themselves to the consequences of their choice (by, for example, having their views published), show that when we have to justify our reasoning, we disengage it from our prior motivation and suddenly start thinking much more like scientists. We consider more alternatives, admit more complexity, devise better tests for determining cause and effect. We fall less often into what's provocatively called the *fundamental attribution error*: assuming that another does what *he* does because that's just his nature, whereas *I* respond appropriately to the situation. The most powerful technique for avoiding motivated reasoning turns out to be imagining how you would argue the other side.

Explaining yourself, admitting other possibilities, thinking like the other fellow—these take us a long way from the flow experience, but they bring the world as a whole much closer to safety. For example, simply asking obstetricians in one hospital to explain voluntarily, without sanction, why they decided to perform a cesarean section rather than a natural childbirth, saw cesarean rates drop from nearly 1 in 4 to 1 in 10, avoiding overuse of a serious surgical procedure with no increased danger to mothers or children.[35] Asking a sample of gay men to record their reason each time they had unprotected sex—even if the reason was "the heat of the moment"—greatly increased condom use.[36]

Despite the medieval paintings, Error is not a single being, born of sin, and enemy of Man. Error, unpersonified, is a part of our thinking process—an ally, if a dangerous one, in understanding and controlling the world. Once we know its taxonomy, from slips to motivated reasoning, we can design our way out of some of it.

We can, for instance, acknowledge the way objects *ask to be used*; handles invite the hand, big red buttons call out for pressing. The brain's dorsal visual system is assessing such objects for likely fit even as its ventral system reads the signs saying "Don't" (and thus we pull the door marked "Push" if the doorknob seems more naturally pullable).[37] A momentary loss of self-restraint—through, for example, fatigue—can allow the body to carry through automatically what the mind has only imagined . . . another reason to store all weapons well out of reach.

We can remember how dull tasks truly numb the mind, so don't expect a bored person to become alert just because the situation has changed.[38] The more familiar a routine, the more we daydream during it—which may be desirable for artists or mathematicians, but is clearly less so for machine tool operators.[39] Our doomed attempts to eliminate *all* risk from a situation may simply make the remaining risks more likely. This is one reason many traffic safety experts now advocate *removing* road signs, boundaries, and guidelines: to transfer responsibility back from the system to the individual and keep the driver's self-protecting mind awake.[40]

We can avoid confusion in this screen- and control-panel-based world through a knowledge of the *Simon effect*: people are much slower at responding to a stimulus if it is on the other side of their visual field from where they must respond. So cockpits now put engine gauges right next to engine controls; well-laid-out computer dialog boxes have "Yes" on the right, near the Enter key, and "No" on the left, near Esc. Similarly, we can exploit the brain's special alertness to the unusual. In the still mostly male world of military aviation, the warning voice of the automatic flight control system is female; pilots call her "Bitchin' Betty"—although she is actually a very nice Alabama lady named Erica. Checklists and standard operating procedures, while dispiriting and time-consuming, organize complex tasks, distribute responsibility, and prevent excesses of self-belief. "Near-miss" reporting prevents experience from dividing between the mundane and the disastrous, admitting that there remains an element of the random in all performance.

The most important element in safety theory, though, is the proper use of other people—and here we return to Naomi the non-sculptor. Gerd Gigerenzer, of the Max Planck Institute for Human Development in Berlin, points out that we are being slightly hard on ourselves in describing our illusions about Naomi as errors; even Stephen Jay Gould, certainly no chump, insisted that "a little homunculus in my head continues to jump up and down, shouting at me, 'but she *can't* just be a bank teller: read the description.'"[41] Gigerenzer's argument is that we are not designed to think logically, but instead to *use all available information*. As with puzzles, riddles, and detective stories, we are aroused by the particular pattern of facts we are given and are unhappy at consigning any one of them to irrelevance: the writer who relies too often on red herrings soon has no readers.

If, however, someone else—not you, me, or Stephen Jay Gould, all of us confused by this conflicting information—had simply asked, "Is someone now leaving college more likely to become a sculptor or a bank teller?" that person would have been useful to have around when we guessed Naomi's career. Modern safety practice attempts to provide that person: a co-worker with similar but not identical information and expertise. First officers have different knowledge from captains'; OT nurses different memories from surgeons'. Simply expecting to take from each according to his abilities is not enough; we need methods to combine them so that the holes in their skills will not, we hope, line up.

In aviation, this practice is called *cockpit resource management*, or CRM—a clunky name for purposeful conversation. It involves separating out and sharing tasks in ways that ensure at least two sets of eyes or ears will be available in situations where one could err fatally. For example, many airlines now have the first officer, not the captain, fly most landings—because handling the plane is actually *less* critical than maintaining accuracy in air-traffic communication and overseeing aircraft systems. Moreover, if the first officer fouls up, the captain can legitimately take over; doing things the other way might smack of mutiny.

The foundation story of CRM is United 232, a DC-10 flying out of Denver toward Chicago on July 19, 1989.[42] An hour into the flight, the Swiss cheese of errors in the plane's manufacture and maintenance suddenly aligned: the engine mounted at the base of its vertical stabilizer shattered, peppering the tail section with shrapnel, and, by dreadful mischance, severing not just the main hydraulic control system but also its two backup systems. The cockpit lost all conventional means of influencing the plane's direction or altitude.

It was a deeply improbable accident, but, as it happens, very similar to one that had caused the crash of a Japan Air Lines flight four years before, in which a fully laden 747 lost its vertical stabilizer and flew, doomed and aimless, for half an hour (the passengers writing their farewell messages) before blundering into Mount Osutaka. It remains the worst single-aircraft accident in history, and it had a morbid but professional interest for Dennis Fitch, a United pilot and flight instructor—who just happened to be a company passenger on Flight 232. He knew that it was possible, in theory, to steer a plane by varying power between its two engines; he'd been trying this out on United's DC-10 simulator. So when the pilots' predicament became clear, Fitch headed up to the cockpit to offer his help.

The recording of the next thirty minutes reveals Fitch, Captain Al Haynes, the rest of the crew, and the approach controllers at Sioux City (the nearest airport) behaving under pressure as we all might wish we could. They are *interested*: "This thing seems to want to go right more than it wants to go left, doesn't it?" They are *adaptable*: "If you can't make the airport, sir, there's an interstate that runs north to south." They pass on significant information: "Anything above two ten knots is gonna give you a nose-up moment." And they are funny: "You're cleared to land on any runway." "Roger. You want to be particular and make it *a* runway?"

There's neither bravado nor pathos, nor, despite the captain's authority, a sense of command and obedience—they sound like a group of friends assembling a jigsaw puzzle. In the background, though, we hear the old authentic voice of Standard Operating Procedure as the airline's maintenance department carries on a separate radio conversation with the flight engineer. At the beginning of the emergency, Maintenance asks: "Is all hydraulic quantity gone?" "Yes." Then, twenty minutes later: "One more time: no hydraulic quantity, is that correct?" "Affirmative, affirmative, *affirmative*." In between are helpful messages like "I'll get your flight manual," "I've got operational engineering on its way over here," and "I've got people out looking for more information," and once more, "You have no control over the aircraft, is that correct?" It's an experience familiar to anyone who has called customer service—but this customer needed something more.

In the end, the remarkable collective effort in cockpit and tower nursed the plane down to Sioux City airport. It was not a perfect happy ending: the plane broke apart when its right wingtip touched the runway and more than a third of those on board died. Even so, had the men in the cockpit not managed their extraordinary feat of teamwork, *all* Flight

232's passengers would have been doomed. This harrowing episode showed how a structured informality can draw out talents, letting a group overcome an emergency that would have been too demanding for any individual and too unpredictable for any set procedure.

People very quickly realized that the world is full of cockpits: small isolated groups of differentially skilled experts who can, after years of routine work, find themselves suddenly faced with an unimagined emergency. CRM has moved, therefore, from the airplane cockpit to that other arena of no second chances: the operating theater. A 1999 report from the National Institutes of Health revealed that up to ninety-eight thousand U.S. patients die each year from preventable medical errors.[43] Dr. Stephen B. Smith of the Nebraska Medical Center says: "We're where the airline industry was 30 years ago. The culture in the operating room has always been the surgeon as the captain at the controls with a crew of anesthesiologists, nurses and techs hinting at problems and hoping they will be addressed. We need to change the culture so communication is more organized, regimented and collaborative, like what you find now in the cockpit of an airplane."[44] No longer, therefore, should theater nurses have to put their hands all over the instruments, rendering them nonsterile, to stop a surgeon they believe is about to make an error.

Of course, medicine is significantly more complex and less repeatable than aviation, and the fear of lawsuits makes no-fault incident reporting—an essential contributor to aviation's improved safety record—a difficult area for doctors. Even so, more hospitals are using the same routines of briefings and debriefings, checklists, and mandatory safety queries and responses that have become the norm in flying—and suffering fewer malpractice cases as a result. "I'm seeing errors caught every day," said Dr. Timothy J. Dowd, chairman of anesthesiology at Vassar Brothers Medical Center. "Even the most curmudgeonly surgeon has to admit this is a better way."

The point is that safety is human and must take account of human qualities. We will never make the world a safer place by sticking on more labels that say "WARNING: Pastry filling may be hot when heated," or "NOTE: Superman cape does not enable wearer to fly." If we encumber every routine with legalistic safety features, it will simply tempt people to disable them in pursuit of flow, even to commit extravagantly risky acts in protest at pettifogging regulations. Since error is intrinsically human, the pursuit of safety must also recognize and reward some subtle human qualities: conscientiousness, companionship, pride in good work, respect given and earned. These are not on the checklist, but every successful program—on land, sea, or air—depends on their help. A "safety culture" *is* a culture, not a rule book.

THINKING PROBABILISTICALLY: THE RULERS OF TANALAND

"Good intentions," said Oscar Wilde, "are useless attempts to meddle with the laws of nature." At Dietrich Dörner's lab in the beautiful Franconian town of Bamberg, study participants doggedly attempt to prove Oscar wrong. Dörner's computer simulations of complex systems have all the infuriating qualities of real life: delays and friction, unseen linkages between phenomena, too much information about some things, too little about others. The tasks range from the trivially technical to matters of life and death: from trying to keep the temperature of a cold store constant after the thermostat has broken, to attempting to maintain the subsistence economy of Tanaland, a fictional region of West Africa.[45] In the refrigeration challenge, the chief problem is delay: turning the coolers up or down produces no immediate change in temperature, which makes one likely to leave them on or off for too long, producing an overshoot in temperature, which tempts one to overcompensate, and so on—to exasperated shouting and pounding of keyboards. Tanaland's issues are more complex and interlinked: the needs and breeding rates of its human population, its cattle, sheep, and leopards, the fruit harvest, the supply of fresh water, and all the other fragile qualities that make one place a land of milk and honey and another a wilderness of desolation.

Children like to imagine themselves as kings; adults, too, enjoy reveries of how pleasant life could be "if only I were in charge." Take command of Tanaland, though, and you will soon ditch your dreams of world domination. Despite the study participants' unlimited power to distribute water, provide fertilizer, build hospitals, donate tractors, and generally exercise the despot's best brand of benevolence, most soon found that they were presiding over a Malthusian nightmare. Within ten simulated years their agrarian peoples, now more than doubled in numbers, were starving; the cattle lay dead in the fields; insects ravaged the fruit harvest because the wild animals that ate them had been exterminated as "pests." As disaster gathered its own momentum, the doomed rulers spent less and less of their time thinking about the problem or asking for information and more and more time making decisions, fine-tuning irrelevant details where they felt they had some personal knowledge or control—*fiddling*, though not in Nero's sense—but by now no decision could reverse the damage. Of the twelve temporary kings of Tanaland, only one maintained stable development, where even the leopards did no worse than before; but this was achieved by painstaking, stepwise application of limited, local decisions—not by imperial decrees.

Even in less demanding theoretical environments than Tanaland, many of Dörner's participants never, ever got the idea; the graphs of their attempts to impose order, even on one humble refrigerator, look like the EKG of a heart attack, all spasm and counter-spasm. Some, however, *did* learn to deal with the idiosyncrasies of their assigned systems, to guess the true contours through the ill-fitting data. What set these candidates apart from their error-prone colleagues? They asked more questions—and their questions were more forensic ("is there a way I can find out if the machines are working at full capacity?" rather than "how am I supposed to know what's going on?"). They described their goals in more detail and elaborated their intended methods further. Their plans tended to be conditional, not peremptory, giving the system a chance to respond before wading in with more stimuli.

Most revealing, though, was their choice of language. Managers who used probabilistic terms like *sometimes*, *in general*, *often*, *a bit*, *specifically*, *to a degree*, *questionable*, *on the other hand*, and so on, were successful. Managers who preferred the absolutes—*always*, *never*, *without exception*, *certainly*, *neither*, *only*, and *must*—were not. A surprise, that: all the strong, sure terms that stud the leadership-course lexicon turn out to be a diagnostic mark for *failure*, whether as the product of a small mind's wish to simplify or a basic confusion between acting confident and being competent. Here as elsewhere, we risk falling victim to the fundamental attribution error, believing that if the quality of a skilled person is decisiveness, then acting decisively must somehow confer skill. But this is to forget what *makes* someone expert: openness, avid curiosity, an eye for variation, and a lust for understanding. These are probabilistic, not absolute, virtues.

The instinct that debases questions of judgment into efforts of will is strong and pervasive wherever people are forced to undergo the pain of uncertainty: in business, in politics, and in the military. It is the source of what's called "bull" in the British army: that attempt to banish war's confusion through obsessive cleanliness, severe haircuts, and massed shouting—controlling anything that can be controlled. But, as every country learns in every war, the peacetime officers who are masters of "bull" are usually disastrous when the real fighting starts. Absolutist thinking gives us the myriad (but always seven-point) mantras of *How to Succeed in Business* manuals, but you will notice that the *top* executives are not the ones buying the books.

Bizarrely, the one top job for which people seem to be chosen purely on the basis of the absolutist quality of their speech is also the most potentially hazardous: high political office. Look at the commentary on elections: how candidates are judged on their apparent "commitment" to their "vision"; how they are excoriated for "flip-flopping"; how evidence that they have made conditional or doubting statements—or, indeed, been influenced by experience—is taken as a sign of moral weakness. Yet real statecraft is complicated and ambiguous. Solemn campaign promises all eventually founder on what Harold Macmillan famously described as "events, dear boy—events." Vote for Woodrow Wilson: "He kept us out of war," that is, until he got us into it. Vote for FDR: "Your sons will not be sent into any foreign wars"—until they were. Successful politicians need to be the *opposite* of what we select for—that is, flexible, attentive, light-footed dealmakers—yet we as voters favor the very qualities that, in Dörner's world, lead to disaster. We wish that the world were less complicated and less dangerous than it is—that there were straightforward solutions to our most pressing problems—and so we reward the candidates who present themselves as best suited to govern that simpler, happy land. If they do so dishonestly to gain our votes, it is a shame on them; but if they really *believe* it and we elect them, the shame is on us.

THE TWO COLONELS

You may still be skeptical about whether this difference between absolutist and probabilistic thinking can really have much bearing on life's complexities, so consider the twin stories of Colonel Osipovich and Colonel Petrov, now obscure men whose actions during the month of September 1983 had a profound effect on history.

Gennadi Osipovich was a pilot, flying his Sukhoi-15 fighter from a secret base on Sakhalin Island.[46] It was an area where the Soviet Union and the West played a particularly tense game of cat and mouse, with U.S. intelligence planes constantly probing the fringes of Soviet airspace to test the reactions of its air defense. On the night of September 1, Colonel Osipovich was on temporary duty, having been recalled from a rare vacation. He was living with his fellow pilots in a little shack at the end of the runway, waiting for something to break the monotony of their patrols. And here it was: something big, perhaps an American RC-135 spy plane, that had brazenly flown right across Kamtchatka. As he trailed it, Osipovich was acutely conscious that he had only a few minutes before the target would leave Soviet airspace. He flew to within two hundred meters of it: "I saw two rows of windows and knew this was a Boeing. I knew this was a civilian plane. But for me this meant nothing. It is easy to turn a civilian plane into one for military use."

In the cockpit, the pilots of Korean Air Lines Flight 007—it was indeed a civilian aircraft, en route from New York to Seoul—were chatting about the best place to change currency at the airport, all unaware of the threat that was now accompanying them. They appear also to have had no idea that they were more than two hundred miles off course, having for some reason set their autopilot to navigate based only on a magnetic compass heading.

An interview with Osipovich thirteen years after the event reveals a simple and consistent worldview. For him, the alert to intercept the intruder was "everything: it means that we just have to go up and kill someone." He followed what he viewed as standard procedures in challenging the Boeing—firing his cannon and sending the Soviet friendly-force electronic signal—apparently without considering that he was firing no tracer (so his shells were invisible in the night sky) and that a Western civil airliner would not have a Soviet military friendly-force responder. He could, of course, have called the airliner on civil frequencies, but "how can I talk with him? You must know the language." He might have told his ground controllers it was a Boeing, but "they never asked me." This is the sort of thinking that generates disaster in Dietrich Dörner's lab—as it did here. Osipovich slipped in behind the Boeing, fired two missiles ("thank God, they hit," he recalled), and killed 269 people.

Just about four weeks later, Lieutenant Colonel Stanislav Petrov was at his commander's position in the bunker at Serpukhov-15, the control center for the Soviet Union's early-warning satellite system.[47] It was a very tense time, and not only because of Colonel Osipovich's actions earlier in the month. NATO was just about to start Exercise Able Archer 83, a simulation of a first strike against the Soviet Union so involved and lifelike that Britain's prime minister and Germany's chancellor acted their allotted roles in it, disappearing from public view. The Soviet Union was now led by Yuri Andropov, the ex-head of the KGB, a man who had breathed deep of the paranoia that seeped through that institution; he was already convinced that an exercise would be the pretext for the inevitable Western attack. This presumption led Moscow to discount the fundamental unlikelihood of an attack and read the most sinister implications into the timing and details of the exercise.

As Petrov was settling in for his shift, the alarms suddenly went off: the satellites reported a missile on its way, launched from the United States. Then there was another, and another—five in all. The alarm system was "roaring," Petrov recalled. "For fifteen seconds, we were in a state of shock. We needed to understand: what's next?"

The general staff had already been alerted. On the console in front of Petrov, the red button marked "START" was flashing. Down the telephone came the shouted order to get on with it and do his job. But, he said, "I had a funny feeling in my gut. I didn't want to make a mistake." Petrov reasoned that "when people start a war, they don't start it with five missiles. You can do little damage with five missiles." He took a breath and made his call—that all this was a false alarm.

So it proved: an unusual reflection from sunlight on high clouds had fooled a satellite system that, in the atmosphere of imminent war, had been rushed into service un-debugged. Petrov's wary, conditional style of thought ("we need to understand"; "I didn't want to make a mistake"; "when people start a war . . .") cut across standard procedures and the delegated responsibilities of the command chain—and possibly saved the world. At least for now: how long it remains saved depends on how many people can think like him.

CHAPTER V

One of Us

Us is different from Me; Them is not the same as You—or even Y'all. We reason differently when we reason collectively. In part, this is because we have an innate confusion between interior and exterior reality. But it's more than that: we share some basic desire to assist with what we think is the plan of our gang. "Our gang," though, is a remarkably limited group: we easily know who is Us in our social and work lives, but we wildly oversimplify when we think about people in general. Once beyond the local, our behavior is shaped by imperatives that aim more for swiftness than accuracy. Ultimately, though, the issue is upbringing: we believe the people we have been brought up with are reasonable and others aren't. The way out is to attribute complexity to others as well as to ourselves.

Oh, the Protestants hate the Catholics
and the Catholics hate the Protestants
and the Hindus hate the Moslems
and everybody hates the Jews . . .
—TOM LEHRER, "National Brotherhood Week,"
That Was the Year That Was, 1965

THEY even had a photograph taken of it. The intersection of Seventeenth and Dodge streets in central Omaha is nothing remarkable today: two office towers, a shopping mall, and a parking garage flank the junction. The Woodmen Tower, tabernacle of insurance, looms a block to the south. Yet on September 28, 1919, this spot saw the climax of an evening of shameful mass savagery.

Men and women, glassy-eyed and grinning, stand in a celebratory circle around their work: the carbonized corpse of Will Brown, whom they had hanged, shot "hundreds of times," dragged around downtown with a rope and then burned, ineffectively, at this bland crossroads. The story behind the deed was a depressingly common one in a summer of lynchings across America: an atmosphere of tension as local white people lost their wartime boom jobs and black people moved up from the depressed rural South; a disreputable newspaper with a political agenda, publicizing "Negro lawlessness" in hopes of discrediting the city authorities; an accusation of sexual assault by a white woman against a black man—and finally, a mob that grew from fifty to thousands in the course of a dull Sunday afternoon.

The point about the Omaha incident isn't just the evil it represents, but the *stupidity*. Yes, the rioters eventually achieved their purpose—killing, as so often, an entirely innocent man—but in the six hours preceding they had also killed one of their own ringleaders and a passerby, half-hanged their mayor, released all the criminals from jail, burned down their own new courthouse and got their town put under martial law, with machine guns posted at downtown corners and a general curfew. To cap it off, there was the photograph—which allowed U.S. Army authorities to identify the principal delinquents and round them up quickly. If race hatred is the desire to gain parity in revenge for injury, the white rioters had put themselves deeply into the loss column without any help from their imagined enemies.

Seventy years later. Another square, another crowd. The gunfire is similar, as is the confused shouting, but something entirely different is going on. In the gray December cold of Bucharest, the mob is intent, not on killing, but on offering itself to be killed. The people are unarmed, surrounded by wavering soldiers, and behind them, the shadowy sniper teams of a paranoid dictator. They have assembled . . . they're not sure why—both government and opposition had called out "the masses." But once assembled, they have a purpose: to bring years of anxiety to resolution; to stand and be counted; to show that the group can be braver than any individual.

That day in 1990, the crowds in Bucharest chanted, "We are the people"; not this people or that—*the* people. In a country so worm-eaten with suspicion that a single Kent cigarette was the price of a betrayal—that, as the journalist Rudolf Kamla put it, "I woke in fear that my wife might have learned what I had dreamt"—strangers now put all they believed in, all they hoped for, right out in the open. They sang, they recited, they wept, drank, and kissed, and—for this was Romania—they smoked, furiously, treasuring each acrid lungful as their last. The soldiers gradually switched over to the demonstrators' side; groups of schoolchildren stood among the tanks as they exchanged fire with secret-police units barricaded in public buildings. Photographs from that heady week show people spontaneously organizing themselves into revolutionary tableaux far more convincing than Delacroix's: a truck, loaded with grandmothers and enveloped in flags, barreling down the shot-riddled boulevard; a bold young pastry chef, staunchly upright, distributing honey cakes to militiamen lying pinned on a corner by sniper fire; a lovely girl tying a rebel brassard to the sleeve of a shy conscript—all cliché, because all eternal and universal.[1] Romanians were simultaneously celebrating and fighting for the shared humanity denied to them by a tyrant in the name of some abstract Socialist brotherhood. Whatever reasons a crowd has for its behavior, Reason itself has little to do with it. History is stuffed with examples of temporary collective lunacy, moments after which the participants must have wondered, separately, "What was I thinking?" The pagans of Ephesus, unnerved by the apostle Paul, chanted in angry chorus, "Great is Diana of the Ephesians" for two straight hours—before breaking up and going home. The Nika riots of Byzantium started with an argument about chariot racing and ended with thirty thousand dead and half the city in ruins. An eighteenth-century Austrian army managed to decimate itself in a single night of confused running battle when the nearest enemy was thirty miles away.[2] Today, we shout from grandstands things we would blush to murmur at home. We succumb to common rumors we would never believe individually—as when, in 2003, the city of Khartoum was overrun by panic over "penis-melting Zionist robot combs."[3] We are collectively fickle and mutable: Freud's nephew Edward Bernays invented the field of public relations in part because he feared the dangerous, libidinal energy that drives the crowd; it needed to be channeled into appropriate behavior through the mass production of shared ideas. From the Great Fear of 1789 to the Summer of Love, from Nuremberg to the Velvet Revolution, the crowd is the domain of license and terror, loyalty and sacrifice.

Is collective irrationality really inevitable? Are these historical examples simply that: *history*, where a particular time and place leaves its unique print on events? Or are there fundamental ways in which joining a group disables our individual ability to perceive and judge?

In 1954, twenty-two eleven-year-old white Protestant boys gathered for a special summer camp at Robbers Cave State Park near Wilburton, Oklahoma.[4] They were not to know that they were actually experimental subjects, nor that the counselors were psychology graduate students and the kindly janitor was really the lead professor, Muzafer Sherif. All they knew was that they were going to meet a bunch of new kids and that it might be fun or it might not.

Sherif divided the boys randomly into two groups, which were initially kept well apart from each other. The boys chose their group names (one was the Eagles, the other the Rattlers), designed their group flags, and went out on activities within their units; they made friends. At that point, the researchers arranged a series of competitive games between the groups, with desirable prizes (including, in those less anxious days, four-bladed camping knives). Almost immediately, a bad-tempered rivalry arose between the Rattlers and the Eagles; there were accusations of cheating, name-calling, surreptitious shoving, refusal to eat together. Things escalated when the Eagles burned the Rattlers' flag and the Rattlers raided the Eagles' cabin, stealing their prizes.

Sherif established a cooling-off separation and then tested to see if shared pleasurable events, like cookouts or fireworks, could bring the boys together—but no: these simply became an occasion for more posturing, insults, and food-throwing. Instead, the catalyst for peace turned out to be shared threat and shared work: when the camp's water supply appeared to have been blocked by "vandals," Eagles and Rattlers assembled as a single group to inspect the pipeline, suggest solutions, and tackle the problem. When a supply truck conveniently broke down just outside the main gate, they joined forces to push it in. At the end of camp, the once-feuding groups asked specifically if they could go *together* on the bus back to town.

Oklahoma fifth graders are not necessarily the pattern for all humankind, but the experience of Robbers Cave looks a lot like some historically familiar events: people sort themselves easily into Us and Them; they unjustifiably favor the former and disparage the latter; and only a larger challenge—not a shared reward—can overcome this prejudice. All too true and all too human. The puzzle is *why*.

FROM ME TO US: HOW THE BRAIN LOOKS OUT

As every magazine publisher knows, people are attracted to faces: their own, the well-known, and the beautiful. Infants only a few hours old distinguish and fix on faces, out of all this dazzling, roaring new world—it's as if they knew that, once they outgrow their temporary helplessness, they will have to negotiate a path around the moods and characters of their families. Mothers and babies spend a lot of time imitating each other's expressions, and it seems time well spent, both as training for conversation and as a diagnostic test: one early sign that a child may not be developing normally is a lack of interest in the traditional mimic games of Hee-hee, Boo-hoo, Uh-oh, and Whoops-a-daisy.

Faces can give a baby more subtle information than just whether Grampa is in the mood to play or not—to a degree, they can reveal whether he really *is* Grampa. Darwin noted the similarity of facial expressions in young and old, across the races and indeed between species, and posited a genetic basis for them. In a fascinating recent study from the University of Haifa, this hunch has been shown to be correct: children blind from birth make the same characteristic facial expressions in a given situation as do their relatives.[5] Even more remarkably, a blind man who had been given up for adoption at two days old displays the same suite of expressions as his birth mother, whom he had not met again until he was eighteen. It is not just the family nose that sets us apart; it can be the family smirk, scowl, or leer. Such signs of kinship can have a powerful effect on behavior: anthropologists have noticed the compulsion in many cultures for the mother's relatives to point out to the father of a newborn how much the baby resembles him—attempting, in effect, to secure a safe-conduct for the child from its most important potential protector . . . or enemy.[6] In a study where subjects were shown paired photographs of faces, one of which was subtly manipulated to resemble their own, they tended to rate the "related" face as more trustworthy but less sexually attractive—a response, the researchers suggest, that supports the impetus to cooperate within the family but not to inbreed.[7] Had Oedipus been a little better with faces, he could have saved everyone a lot of trouble.

What, besides kinship, are we looking for in faces? How is it that we can parse in a moment the nuances of emotional byplay that Henry James takes whole chapters to describe? What makes this awareness so intuitive and makes humans, as the psychologist Giacomo Rizzolatti describes them, "an exquisitely social species"? He and his colleagues at the University of Parma came across the beginning of an answer when they were looking at how the macaque brain plans motion, recording the firing of individual neurons in the pre-motor area as the monkey reached for and ate a peanut.[8] In the course of the study, they found that certain neurons were firing, not just when the monkey reached for food, but also when it saw a *researcher* reach for food. These *mirror neurons* bridge the philosophical gap between internal and external reality: whatever may be the ontological truth, the same neural pathways map a primate's own actions and the actions it observes. There is an implicit assumption of equivalence between self and other—*I think, therefore you are.*

Further studies have identified mirror neurons (or more properly, mirror networks) in humans. You can feel them at work right now, as I tell you exactly how I once got this terrible paper cut (*stop, stop!*) on my eyeball (*auggh!*), or if you try to play the trombone while watching someone chew on a lemon. From aping motion, these systems move us toward empathy.[9] They make us not just squeamish but tenderhearted; awake us not just to others' rage and fear but also to such subtleties as disappointment, vain hope, ennui, and—for extra credit—irony. They make us spectators feel we are playing, in a shadowy way, the sports we watch—even just *thinking* about exercise can slightly improve strength and muscle tone.[10] Doing, feeling, observing, and imagining may be separate experiences, but they are all vehicles using the same highway.

What about those for whom the traffic is mostly one-way—who lack the sense that others feel and plan as they do, or that other people's emotions matter? Certainly, the information from at least one comparative brain-imaging study confirms what most would intuitively believe: men are shorter on mirror neurons than women.[11] We all know someone whose tactlessness seems a genuine social disability, not just a character trait. In its more extreme form, this empathy-blindness can be a symptom of autistic spectrum disorders—and though the neuroscience is still in its early stages, there is evidence that the severity of autism can be correlated with a thinness of cortical areas related to mirror networks.[12] When combined with the gender-linked results, this tends to support Simon Baron-Cohen's theory that the autistic mind is simply the extreme male mind without any social extras: optimized for track racing, but not street legal.

The capacity for empathy, from tingling when others are pricked to weeping at others' weddings, has many implications for how we approach the world. And it is not just us: mice will wriggle in sympathy with a cage-mate's stomachache.[13] Chimpanzees can sense, not just the emotions of other chimpanzees, but something about their goals—especially in competitive situations involving food ("he *will want to* take my banana; I'll retrieve it *when he won't* notice").[14] Capuchin monkeys can tell whether a human is fiddling with a container because he thinks there's food inside or because he's just messing around.[15] Assumptions about goals are the first step toward society: if I think you want what I want, we could work together to get it. Chimpanzees certainly operate on this principle. Not only will they cooperate with humans to get food out of a puzzle box; they will also recruit help for tasks based both on the complexity of the job and the past competence of the potential collaborator.[16] Such results produce a combined sense of "*amazing!*" and "*well, of course*": apes make the same decisions we do when building a team or a gang. That's pretty remarkable, but then, why else *have* a team or gang unless there is some task beyond individual competence? They and we are social out of necessity—we want the banana or its equivalent, so we club together. Or is there something more to it?

At the Max Planck Institute in Leipzig, in a cheerless little conference room, a researcher is hanging out sheets to dry.[17] As it happens, he's not doing a very good job—but you feel for him, with his plump earnest face, cloud of fluffy hair, and little oval Schubert glasses. Some poor postdoc, confined to the lab by his slave-driving supervisor, having to do his laundry at work . . . Is this what is going through the mind of the eighteen-month-old baby, sitting there in his mother's lap, intently watching the bumbling student? Probably not. But when the man drops a clothespin where he can't reach it and lets out a despairing "*oh!*" the infant heaves himself to his feet and toddles laboriously across the room to pick it up and hand it over. Unlike the ape experiments, there is no yummy reward involved, no approval from elders, no stimulus except another's need. Human babies apparently want to help from a pure desire to be involved; they know what help is needed, not because they understand what the final goal is, but because they empathize with another's frustration and disappointment.

Psychologists have assembled these various phenomena of anticipated desires and guessed intentions into a complete hypothesis: the so-called *theory of mind,* which postulates that human intelligence differs from that of all other animals in that we alone can imagine another's beliefs—including, or especially, *false* beliefs. One telling bit of evidence for this being a uniquely human capacity is that even humans come to it pretty late. The classical false-belief experiment goes like this: you see two people (or dolls, or cartoon characters) playing with a toy in a room.[18] They put the toy in a box, after which one of them, Sally, leaves; the other, Anne, then takes out the toy and hides it somewhere else. Sally comes back and the question is, "Where does Sally think the toy is?" Most three- and four-year-olds assume that Sally knows what the rest of us know—the toy has moved. But five-year-olds can imagine her *not* knowing: the power of fiction has suddenly been revealed to them (this is also the age when they become much better liars). Meanwhile, younger children will consistently offer crackers rather than raw broccoli, even to those researchers who have bravely pretended that they hate crackers and dote on broccoli—because *everyone* knows broccoli is revolting.[19]

To imagine a mind that prefers broccoli to crackers is a demanding but useful mental stretch, because it opens an entirely new method for understanding the world. Although learning from experience—personal science—may seem the simplest route to gaining knowledge and expertise, that's only because it's the easiest method to describe. As every scientist knows, understanding what your experiment is really telling you, or even whether it *can* tell you anything, is far from straightforward. As for "I've seen it with my own eyes," we all know how trustworthy *that* evidence can be. Our senses are fine for the here and now, but to develop useful truths about the world in general, we need the ability to imagine and reproduce new, alien experience—including the experience of being wrong about something. "What if *I* were Sally?" is always a question worth asking.

HOW BIG IS YOUR GANG?

To whom, then, are we willing to extend this theory of mind? Despite its roots in the family, our sense of "us" seems surprisingly flexible, not necessarily requiring birth kinship or even a shared species. Children have been raised in the society of dogs, monkeys, wolves, and gazelles, taking on the habits of their substitute families—often to the detriment of their uniquely human qualities, such as the ability to acquire spoken language. Yet, at the same time, the attributes of belonging can show remarkable persistence. As immigrants, we preserve the accents, tastes, and attitudes of "our own kind" even after a lifetime spent among others; as the British-Hungarian producer Alexander Korda admonished his nearly drowned brother, "A cry for help should always be in your native language."[20] So, whether you were raised by wolves or by Hungarians, you will have a feeling that the world divides into us and the others—that there is a core community to which you truly belong, separate from the mass.

How big is this community? In theory, it can range from the embracing *ummah*, the worldwide kinship of all Muslim believers, to the sparse Yankee idea of home as "where, when you have to go there, they have to take you in."[21] We can feel a strong, inarticulate temporary community with, say, everyone at this wonderful concert, this righteous protest, this three-pointer-at-the-buzzer victory, or even this cozy bar and grill—but such vague communities have their limits. You would not, for example, lend your outboard motor to some stranger just because he was a fellow citizen of Red Sox Nation. How big, then, is the real community: the people you would feed if they hunger and whom you would expect to feed you? Roughly 150—and, within that, a maximum of 12 whom you genuinely love as you love yourself.

These surprisingly narrow estimates come from the work of Robin Dunbar. For the past twenty years, he and his colleagues have been looking at the social networks of human and other primates and have come up with a fascinating correlation: the larger the neocortex in a primate brain, the larger the typical social group for its species (adding credence to the stories of children raised by wolves, the same correlation seems to apply to other carnivores).[22] It appears that the maintenance of complex social relationships—keeping track of status, mating, food distribution, grooming—makes heavy demands on limited brain resources. There are, however, distinct advantages in having such detailed social information—in knowing, not just that you can count on your friends, but *which* friend you can count on for *what*. This means that cortical volume imposes an upper limit to the number of individuals we primates can know *as* individuals. With humans, the curve reaches its maximum of 150, 12 of whom are our "grooming clique," with whom we share the kind of intimacy where even long silences feel relaxed. You might notice the power of this maximum when you catch yourself saying, "Why don't we ever see the So-and-sos anymore?" Old acquaintance is forgot because new acquaintance takes its place. When we bring it to mind, we do so less to renew it than to indulge a bittersweet nostalgia for the selves we once were.

A community of 150 fits well with average hunting group and hamlet sizes for pre-technological cultures, but Dunbar has also tested his prediction in our more complex society, using one of those elegant data sets that subjects generate themselves: Christmas card lists.[23] Sure enough, the circle of people deemed important enough to contact at least once a year—relatives, old friends, work colleagues, people one can't stand but feels obliged to—averages around 130, with a maximum network size of 153.* The limit is robust: anyone with the ability to care *individually* about more than 150 acquaintances is a prodigy and should probably become a diplomat.

Or a gossip columnist—for the glue that holds all primate networks together is the scandalous chat that even priests and professors retail in private. Power, sex, conflict, children—and tying all those

* The study passes over those prominent men and women for whom a bulging Christmas card list is a professional responsibility: George H. W. Bush's is rumored to be in the thousands.

neatly together, adultery—are the most compelling topics in social discourse. In one revealing memory experiment, people playing the "telephone game," where phrases are whispered down a line of players, lost far less information about third parties when it was couched in terms of gossip rather than more abstract, general facts.[24] Gossip isn't just fun; it conveys a précis of survival-fitness that would be essential to know in more primitive times: who's fertile, who's first at the kill, who's nursing a grudge, whose baby is this?

Our problem now is that times are less primitive, and the rules of the primate band sit awkwardly with a life in which we are both more crowded and more alone. Our helpless receptivity to gossip puts us at the mercy of soap opera and celebrity culture, where we know more about the intimate life of Britney Spears than about our own nieces. We judge the strangers from whom we require purely abstract services (political candidates, business leaders) as if they were friends and troop mates, using criteria a baboon would find appropriate: hair thickness, waist-to-hip ratio, monogamy—none of which is necessarily relevant. And this closeness, of course, only works in one direction: the face on the screen will never be smiling just for you. If a significant proportion of your 150 known people are celebrities or fictitious characters, it removes the potential for that number of more fulfilling, genuinely reciprocal relationships. A child brought up with media friends is essentially lonelier than one whose friends are imaginary.

Once we are grown up, our 150 acquaintances will also include everyone from work, people whom we no more chose than we did our parents and whose right to our loyalty is, theoretically, pretty tenuous. Yet we give it: in a *Wall Street Journal* poll, 71 percent of employees described themselves as loyal to their companies (34 percent as "extremely" loyal), despite 34 percent describing their organization's leadership as no more than "mildly" trustworthy.[25] Work is necessarily an exchange of individual skills for shared goals—which means that, like the helpful Leipzig babies, we are predisposed to bond with colleagues, to seek to help them and to deserve their admiration. They are firmly in our primate social network whether we want them there or not; it's what makes losing a job so painful, even if it wasn't a job you much liked when you had it: you've been exiled from the band.

The worst effect of our limited social perception, though, is what it does to our view of those outside it: the 5,999,999,850 people we cannot see as individuals and must group together into collective nouns—that is, as stereotypes.

FACING UP TO THEM

I don't know how you feel about the Pirese. I'm pretty tolerant. I enjoy a mixed neighborhood, I believe religion is a private matter, and as for immigration—well, like it or not, we need the skills. But these *Pirese* . . . I kind of draw the line there. I've got kids. Any Pirese can just keep moving on, as far as I'm concerned.

And I'm not alone: a 2006 Tárki Social Research Institute poll in Hungary revealed that Pirese refugees were hated even more than Romanians, Russians, Chinese, or Arabs.[26] It wasn't so much the things they had done—they don't leave much of a record—but what they *were*: dark, said some respondents; ugly; possessed of the evil eye; known to mix blood into their beer. And the worst of it is, they don't exist—they were just included in the questionnaire as a statistical control. Not even *existing*: typical sneaky Pirese behavior.

Rudyard Kipling, better psychologist than he gets credit for, sang:

Father, Mother, and Me,
Sister and Auntie say . . .
All nice people, like Us, are We,
And every one else is They.[27]

True enough, but why should We *hate* Them, rather than just treating Them as an inevitable but unexceptional part of the world, like trees or animals? Part of the answer stems from those sharing games we saw being played in Professor Fehr's Zurich lab in chapter II: the most precious commodity for a social animal is *trust*, because it allows for division of labor. A primate band whose members can trust each other can also share food, or at least the knowledge of where food is abundant. Hunters can concentrate on hunting in the confidence that gatherers are gathering. The band does better than any individual or mother-child unit could do. And while it may not be immediately obvious why this helps any particular gene make it to the next generation, it buys time for *all* the group's genes: if everyone lives longer, it increases the chance that, say, today's foolhardy juvenile might survive to breed as tomorrow's experienced troop leader. If "Father, Mother, and Me" believe that even Mad Cousin Anne is one of Us, she and we all have better survival chances—assuming she reciprocates our trust.

There's the rub: trust by its nature opens the door to cheating, which is surely why the participants in Professor Fehr's games stepped in to punish freeloaders with such ferocity. If our most significant adaptive advantage depends on a behavior that itself makes us vulnerable, we will be preternaturally aware of any threat from that direction—especially in a world with, as we always believe, too few good things to go around. Indeed, the economist Samuel Bowles suggests that our altruistic behavior to those we consider in-group is a direct *result* of lethal competition for resources among early human groups—that Us first appeared as the response to a threat from Them.[28] Humans differ fundamentally from other social animals, Bowles maintains. Our ability to pass on information over time, our traditional monogamy, and our complex cultures tend to bind together our individual destinies into group fates, while our intelligence and persistence make our conflicts particularly deadly—every human quarrel contains the germ of a genocide. Threat therefore generates solidarity: the warlike Yanomamö of the Amazon share food outside their immediate family *because* they need joint protection against neighboring enemy villages.[29] Highland peasants, whether in Greece, Scotland, or Afghanistan, devise complex and sacred rules of hospitality *because* of the prevailing culture of blood feud and mutual plunder. Even in the most comfortable modern life, there is that tickle of potential conflict for resources: if, before the sale opens, you save places in the line for your family or friends, you're being a good person, freeing them from the boredom of waiting—but the lady in front of you doing the same thing is *cheating*, holding you back in favor of her deadbeat companions who can't even be bothered to show up. We bond against as much as with. Every people instinctively heeds Benjamin Franklin's advice that "we must all hang together—or assuredly we shall all hang separately."

So how do we know who They are? It's not always easy. Ancient Israelites asked refugees to pronounce the word *shibboleth*; the Danish resistance in World War II preferred *rødgrød med fløde* (red berry pudding with cream); roving Croat and Serb terror gangs checked the manner in which local children crossed themselves—Orthodox do it right to left, Catholics left to right.[30] At the most basic level, though, our feeling of otherness is taken in with our first nourishment and is deeply intermingled with our sense of disgust.

Grubs, sheep's eyes, raw fish, tripe, haggis, pork, beef—*They* eat revolting things. Dietary preferences are set very young (if you want your children to be gastronomically liberal, you need to feed them small amounts of varied foods from an early age) and these preferences quickly impose mental boundaries between the acceptable and the disgusting. Even if we don't actually *see* people eating horrible stuff, they can remind us of it simply by their bodily presence. At various times, the Japanese have complained that Dutchmen smell like sour milk; the Chinese that Europeans smell of cold meat. Distinctive regional cuisine can certainly create a shared crowd odor, imperceptible to locals but obvious to the visitor: cumin in India, garlic in Korea, and that spicy, dusty scent like sun-warmed adobe in parts of Africa, ascribed by keen noses to a diet of cornmeal. The potential for disgust from the ethnic diversity of personal smell might explain why, in the melting pot of America, there are so many ads for mouthwash and deodorant.

Jonathan Haidt and his colleagues have looked at the role of disgust across cultures and have found that while its *physical* sources are the same everywhere (feces, corpses, and so on), its moral and intellectual triggers differ markedly.[31] Americans say they are "disgusted" by senseless violence and racist attitudes; Japanese by day-to-day impoliteness and their own personal failings; Israelis by lying politicians and naked fat people on the beach; ancient Greeks by shamelessness; and the Hopi (at least the six Haidt interviewed) by cosmic imbalance. Varied sources, but one, queasy reaction—which means that when cultures collide, it is almost impossible to forgive the initially inadvertent moral slights that each one gives to the other, because the reaction they produce is not moral, but physical: *yuck*.

Disgust is mediated in the brain by the insula, that deep-set unchanging primate structure that apparently governs gut feeling and the visceral emotions, such as fear and anger. Some research suggests that each separate unpleasant feeling will call up, like a witch's familiar, its own distinctive style of hating the stranger.[32] Disgust is manifested through the language of contamination ("they're spreading everywhere"; "would you want your sister to marry one?"); fear by talk of shadowy practices and ritual savagery. Do you feel contempt, or pity? You will characterize the Other as stupid or comically incompetent (as the English did the Irish and the Indians). Is envy your primary emotion? Then the Other appears frigidly aloof and inhuman (as the Irish and the Indians saw the English). We build a cultural complex on top of these fundamental emotions, from which we keep a lookout for behavior that will confirm our insula's instant prejudice.

This means that deep-seated mistrust takes a long time to dispel. Americans may say they find racist attitudes disgusting, but they still have them—and not, unfortunately, in places that are easy to get at. A neuroimaging study reported by Alexandra Golby confirms the bad old joke: people from other races all look alike. We are better at distinguishing unknown faces from our own race; specifically, we are more likely to activate the *fusiform face area*, a part of the fusiform gyrus that plays an essential part in recognizing individuals.[33] Interestingly, this same-race bias is stronger among white Americans than black ones, suggesting that the experience of being a minority in the population makes black people more adept at distinguishing majority white faces; white people, by contrast, seem more likely to stop at the categorical identification—race—without passing on to the individual. And it's not just a matter of race *identification*; there is also an intrinsic emotional response. Although combined neuroimaging and conventional psychological testing of white Americans found little evidence of explicit racism (their responses to a standard questionnaire showed, if anything, a pro-black bias), many subjects showed an unconscious anti-black tendency when faced with the task of sorting favorable and unfavorable words along with white and black faces: it took longer to associate the good with the black than with the white.[34] The subjects who scored high on this implicit racism test also showed a higher startle response to black faces, with more activation of the amygdala, a structure that mediates emotional response to remembered, often unpleasant, stimuli. These neural responses disappeared, though, when the black and white faces were familiar to the subject, implying some support for the better class of old joke: "He's not Black, he's Jones."

Does this mean that we are hardwired to be racist? No—as the researchers take pains to point out. What the research *does* show, though, is that exorcizing this shameful imp is more than just an intellectual exercise. It requires being aware of how closely entangled our prejudice can be with our physical sense of well-being—and, like trying a new food, takes an effort of will.

THE OTHER WHO IS MYSELF

Eleanor Roosevelt said, "No one can make you feel inferior without your consent." Like many of Mrs. Roosevelt's observations, this is much harder to live up to than to say.[35] We may be selfish as a species, but we are not very self-aware, or there would not be personal advice columns in every newspaper. Our natural bonds to the group mean we tend to take the opinions of others, good or bad, very much to heart. Thus many a well-stroked chief executive will come to believe he is indeed as mighty as Jehovah, while those who are despised and rejected cannot help but feel they have taken on some taint from all the mud flung at them.

In its extreme form, this appears in the idea of the "usual suspects." If every local theft is blamed on us Gypsies, what have we got to lose in living up to our reputation, ripping off the *gadjo* while entertaining ourselves with trickster tales of daring pilferage? If the police seem to assume that every black youth on the street is either a dealer or a pimp, why not elevate those badges of shame into titles of counter-heroic gangsta achievement? Victims of prejudice can find themselves in a trap, where to assimilate and succeed in wider society seems a surrender to the oppressor and a betrayal of one's group.

The Buraku people of Japan, for instance, differ from the surrounding population *solely* by their places of settlement (plus some revelatory family names), but they are the descendants of a ritually impure caste of leather workers, butchers, and undertakers, known in feudal times as *eta*, or "extreme filth." Official discrimination ended in 1871, although collective titles such as "newly ordinary people" kept the sense of otherness alive. Today, there are probably between two and four million Burakumin in Japan; a majority of other Japanese would still object to their child marrying one, although such marriage is now more frequent. Secret directories of Buraku home addresses have circulated among the personnel departments of major companies to help those "grappling with employment issues." Buraku villages still have lower incomes, higher unemployment, and lower school enrollment figures than the national average. And yet there is a kind of vicious symbiosis between the tradition of discrimination and the potential for criminal gain; recent articles have exposed overlaps in membership between the main Buraku lobbying groups and *yakuza* criminal families—much as the early U.S. Mafia piggybacked on the self-help activities of the Unione Siciliana.[36]

If crime is an extreme form of buying into negative discrimination, are there subtler ways in which we take on the prejudice of those whom we suspect secretly despise us? Indeed there are, and for these we must turn to the unsettling experiments of Claude Steele. Professor Steele, now at Stanford, is himself the product of an interracial marriage and learned early on how much more there is to a racial identity than simply skin color. As an academic in the late 1980s at the University of Michigan (a socially liberal but scholastically demanding school), he was puzzled and upset that black students arriving with SAT scores on a par with those of their white classmates were ending up with lower grades and dropping out in larger numbers. He wondered if this was the result of some internalized prejudice, a specific failure of self-belief he called *stereotype threat.* So he set up a situation in which white and black students with equivalent aptitude had to take a challenging portion of the GRE, an exam for graduate-school applicants. When the subjects were told this was a "test of verbal ability," the black students did markedly less well than the white—but when told it was a "problem-solving" challenge without academic content, both groups did equally well. It seems the black students were being held back, not by a general intimation of inferiority, but by specific beliefs about what some groups are good and bad at.[37]

To test this hypothesis, Steele and his colleagues extended the range of challenges and the groups who undertook them. The results showed a dense web of unconscious prejudice: women did worse in difficult math tests than men—except when told that gender was not expected to affect the scores.[38] White males did less well at math than Asian-Americans, although here the sense was not so much of personal inferiority but of Asian superiority.[39] White golfers told that the game depends on "natural athletic ability" played worse than when they believed it was a matter of strategic thinking.[40] And white students jumped even less high when being assessed by a black coach.[41]

The worst situation, of course, is where the stereotype threat of one group aligns with the self-serving prejudice of another: where pairs of populations fall into a terrible codependency, as destructive and difficult to escape from as an abusive marriage.[42] Caste systems everywhere draw their power and persistence from an unconscious agreement that the world does indeed sort into such categories and that some must be down if others are up. Much of the time, these arrangements are the result of conquest (Aryan and Dravidian, Tutsi and Hutu) or the appropriation of power (Russian serfdom, "benign" colonialism). But people don't actually need any racial, national, or historical excuse: you can generate the same effect in the lab.

The 1971 Zimbardo prison experiment at Stanford has a notoriety in the psychological literature second only to the Milgram electroshock studies of 1963.[43] Its purpose was to investigate whether preexisting character traits influenced behavior in a prison environment—whether the cruelty and stress of prison life was somehow intrinsic to the situation or simply a reflection of the kind of people who were likely to end up as prisoners and prison guards.

From eighteen middle-class white male volunteers, a random nine were designated criminals, "arrested" at their homes and whisked to an improvised but effective prison facility built in the basement of the psychology department. They were fingerprinted, stripped, sprayed with insecticide, and had their heads shaved. The other nine had been designated guards; they got khaki uniforms, billy clubs, whistles, and mirror sunglasses. There was little established procedure: the researchers remained intentionally distant to see what would happen.

What did happen was alarming and depressing: the full panoply of prison pathology in both guards and prisoners—bullying, threats, betrayal, passivity, severe psychological disturbance—all erupting in less than a week, to such an extent that the study was ended early. This disappointed the guards, who had become fully involved in what they saw as the task of keeping the prisoners "under control." Nor was it just the guards: Philip Zimbardo reports of himself that, a few days into the experiment, he was no longer treating prisoner resistance as an interesting experimental result, but as a threat to the integrity of his model prison. He, too, had become a jailer.

It would be easy to conclude from this that power relationships are intrinsically evil, but the truth is more nuanced. It's only empty power that we loathe as tyranny. We seem to respond favorably to power when it has *content*: who's in charge here? How do successful executives operate? What does Mom think? Our intimate, local groups sort without trouble into leaders and led, planners and accomplishers. In our brains, this feeling for status is mediated by serotonin, the neurotransmitter whose lack is implicated in many cases of depression and senseless anger. A landmark study by Michael J. Raleigh in 1984 showed a strong correlation between blood serotonin concentrations and dominance in vervet monkeys. The top male had roughly twice the levels subordinates had (the top professor in the study also had twice the levels of the lowly graduate student): being leader is rewarding. But *becoming* leader is equally so: gaining a dominant place is reflected in rising serotonin levels.[44] Even the trappings and rituals of dominance have a favorable effect; it's the reason behind every medal, mortarboard, and plaque of appreciation, and the explanation for why large companies have so many vice presidents. A higher serotonin level produces the same satisfaction as a higher salary—and need cost no more than the expense of printing new business cards.

Empty, unjustified dominance, though, has a drastic lowering effect on the serotonin levels of the dominated (with a corresponding rise in the stress hormone, cortisol)—perhaps the reason such a high proportion of Zimbardo's prisoners suffered extreme psychological reactions. After all, his subjects were just *kids*: college-age comfortable Californians with no knowledge of the extremes of human experience. They knew no *right* way to behave because they lacked a detailed mental image for the roles they took, and Zimbardo purposefully gave them no instruction. Bleached of content, the situation reduced to its simplest terms: physical compulsion—that sickening knot of rage and fear we remember from long-ago playgrounds. It let loose two of the most powerful and shameful of human emotions: guilt at our own impotence and hatred for those we have wronged.

THE PROFANE MOB

Crowds don't think, they feel. If you have ever been in a riot or a panic, you will remember how little it takes for a powerful emotion to surge through the mass—and how quickly it happens. As Tolstoy claimed, the issue of a great battle can depend simply on whether some anonymous soldier at the head of the column shouts "Hurrah!" or "All is lost!" Our relations with the people beyond our local group of 150 complex, interesting characters may be momentary, but those moments can be highly intense.

This has much to do with the familiar but unconscious play of *social signals*: the nonlinguistic techniques we use to convey our states of mind—confidence, interest, dominance, flexibility. Alexander Pentland studies these cues of voice-tone and posture, and his work confirms how quickly we take in signals and how powerfully they predict the outcome of short-term interactions.[45] If, for instance, you can maintain an engaged, high-energy manner during just the first five minutes of an hourlong pay negotiation, you will end up with a higher award. Show a give-and-take conversational style and you will be more likely to become the "one who knows" in your social network—people will trust you with information. During speed dating, if the woman's expression is active, *both* parties will feel more attracted; if both appear engaged with the conversation, they are more likely to report feelings of friendship than of lust; and if the man mirrors the woman's posture, she will trust him more. All this confirms the well-documented importance of first impressions: people are very quick to make emotional judgments.

Quick, of course, need not always mean correct. Snap impressions are fine for brief encounters but often less useful for the long run. We all know charming cheaters, imposing and handsome airheads, and dynamic leader-types with no sense of direction. Love at first sight is no promise of love forever. The writer Edna O'Brien said ruefully that she was inevitably attracted to tall, good-looking men "who have one common denominator—they must be lurking bastards." Thin-slicing experience doesn't always leave you with the cut you wanted.

Despite this unreliability, social signals retain great power, probably because of their antiquity. Every flock, herd, or school of social animals uses them to generate a collective response to threats or opportunities; thus, when a thousand starlings jink through the sky, their movements seem the product of one mind. Group-living mammals use posture and voice-tone to pass on essential messages about danger, supremacy, and mating status—or all three combined, as when a dominant male chimpanzee gives his formulaic tree-shaking exhibition of rage. He plunges and bounds, and the whole troop immediately starts shrieking, including those who could not possibly have seen the original threat display: the mood-message passes almost instantly through a telegraph of nonlinguistic cues. Similarly among humans, gesture and vocal expression operate on a deeper plane than language, giving our banal conversation most of its actual meaning. They convey the germs of emotional contagion: smile begets smile, sneer leads to insult—"don't be so defensive!" we exclaim, offensively. Logic need have little to do with it: one customer-service representative at the other end of a phone may seem charmingly inept, while another is infuriatingly incompetent—the only difference is *tone*, but tone travels faster and further than sense. So whether we respond to the body language of one person across the table or to the shouts of a mob across the square, we do not necessarily engage the cool reasoning faculties of our prefrontal cortex. We are more likely to draw on the wordless, passionate resources of our old mammalian brain.

Of course, not everyone likes the idea of being related to other mammals. For some, a fear of simian disorder, and hatred of the Other (whoever that may be) can combine into a powerful urge toward discipline for its own sake—toward authoritarianism. At the University of

Manitoba, Bob Altemeyer has made authoritarians his special study.[46] His innocent-appearing agree/disagree questionnaires give his subjects ample opportunity to reveal their inner monologues: "1. The established authorities generally turn out to be right about things, while the radicals and protestors are usually just 'loud mouths' showing off their ignorance . . . 5. It is always better to trust the judgment of the proper authorities in government and religion than to listen to the noisy rabble-rousers in our society who are trying to create doubt in people's minds . . . 8. There is absolutely nothing wrong with nudist camps." (This, last, one suspects, would count as a *strongly disagree.*)

Top scores on Altemeyer's test indicate people with "a high degree of submission to the established, legitimate authorities in their society; high levels of aggression in the name of their authorities; and a high level of conventionalism." He calls them *right-wing authoritarians* (RWAs), but by this he does not imply that they have particular opinions about Keynesian economics or gun ownership. Instead, he means that they show a desire to be among the "right-thinking": a straightness defined more by rigidity than direction.

Indeed, the authoritarian desires to follow authority *wherever* it may lead—and in this respect Nazi storm troopers differ very little from Stalinist shock workers. The RWA sees his role as finding and joining any majority, defending the body politic against dangerous free radicals: the invented subpopulations (brigands, poachers, townies, jitterbugs, crypto-bourgeoisie, left-deviationists) that get the blame whenever there's trouble. In a cunning further study, Altemeyer asked subjects who had high scores on his RWA test whether they would, if the federal government had passed a law outlawing certain religious cults, willingly help hunt down and arrest members of those cults or participate in attacks on their meeting places "if organized by the proper authorities." Why, yes, they would. Then Altemeyer adjusted the target group of this putative federal ban: Communists? Sure. Homosexuals or "unpatriotic" journalists? Certainly. The Ku Klux Klan? Well, OK—if the authorities said so. What about "right-wing authoritarians?" Uh . . . yes. Although their agreement was less emphatic, most RWAs were so keen to side with power that they would be willing to join a posse to persecute *themselves*.[47] You can only imagine what they would do to the Pirese.

I LIKE YOU, BUT NOT YOUR BRAIN

In the more robust days of American politics, James Thurber's great-grandfather had a special way of reasoning with people who did not share his favorable view of Andrew Jackson: he would heave them across the road like a caber. If we are going to discuss all the ways people can't get along, we should include those we hate for their thinking, not just their otherness. Our world is a roiling sink of opinion, from what makes a just society to what makes a perfect martini—so there is obviously no shortage of things we can disagree about.

A good place to start an argument is with a counterexample or a contradiction, and in this case I* happen to know someone who is both. Will Thomas does not share a single political opinion with me: his views are an amalgam of original-meaning constitutionalism, natural-law theory, the charter of the National Rifle Association, and Ludwig Ott's *Fundamentals of Catholic Dogma*. His support for his positions is rigorous, articulate, and learned—and I can't bring myself to agree with any part of them. Yet he is *personally* one of the best men I know: a true friend, a sincere counselor, a devoted husband and father. If I were on the run, his would be the door I would knock on. For his part, he is willing to tolerate my heterodoxy, even assuring me that, in the current state of doctrine, it is unlikely that I am *eternally* damned.

Now, if we cannot agree, how can we be friends? Or if we are friends, why can we not agree? It is not that neither of us has yet come up with the clincher, the all-conquering logical weapon that will convert the unconvertible. We will *never* agree, because our views are ultimately inseparable from our identities.

---

* Michael is speaking here.

Why, though, should opinion become merged with self? Why do people describe themselves as "socialist" or "Orthodox" as a primary identifying label when so few aspects of life actually require such an ideological approach? It may all have to do with early childhood. When we consider how babies learn things, we tend to concentrate on their amazing ability to pick up language or come to conclusions about the physical world. This is a rich and exciting field of study and has provided all sorts of delightful insights, from Chomskian innate grammar to Bayesian probability judgment. But all this is to ignore two facts: children learn much more than they have time to test, and they learn whole scads of stuff that is *impossible* to test. By the age of six, most kids can tell you things about the diplodocus that you never knew; in exchange, you are telling them things (such as "the world is very old"; "the body contains many different vital organs with separate functions"; "God knows all things") that they must take only on your say-so.

This is what Paul Harris at Harvard studies: the great majority of information about the world that children must accept through testimony rather than personal experience.[48] They are not, he would argue, little scientists; they are more like little jurors: they apply critical judgment to the plausibility of statements and the credibility of witnesses. Surprisingly young—sixteen-month-old—children can identify when an adult has said something that is not the case: calling a dog a "cup," for instance. This bothers them; they try to correct it. By age three, they will tend to discount further things said by such an untrustworthy adult.[49] Trustworthy adults, though, become progressively more and more credible as sources of information. So if, say, your aunt Maria is kind and honest, always talks to you like a grown-up, and makes great cookies, you are likely to pay more attention to her views on foreigners, fate, or the afterlife—batty though these might be. Like most childhood acquisitions, her views will stay with you, wrapped into your sense of who you are. They will seem just as true as the physical facts you learned by test and observation.

Now, Will Thomas had wonderful parents: strong, warm, moral people who lived in a small farming town. Their worldview derived from their own pious and hardworking experience and from the particular enclosed history of their place. From a child's perspective, they are *drenched* in credibility, which means I could no more argue Will out of his inherited opinions than I could make him deny his parentage. What he thinks about papal infallibility is a part of the fabric of trust and trustworthiness that binds him to his people (much as my own skeptical instincts bind me to mine). There is no real difference between basic abstract beliefs and family tradition—a fact borne out in Paul Harris's work: children's opinions about whether the brain is the seat of personality, whether the world is round, and whether all species were created by God seem to be firmly in place before they go to school.[50] And if a particular belief stands at the center of family or community life, no subsequent information will change it. Most children of fundamentalist families retain their belief in creation of species right up through adulthood, no matter what they are taught in school; children of nonfundamentalist families shift toward evolutionary explanations as they learn more about fossils.

Such early acquisition means that belief can feel essential to one's identity without having to be logical. Children have neither the time nor the opportunity to test concepts; they have to adopt such quick-and-dirty heuristics as "Dad said it, so I believe it." This means they are stocking their minds with axioms rather than testable propositions: truths outside logic, to which they will have to tailor their later observations and conclusions. It explains why so much of the current debate about religion versus science is essentially irresoluble; as Jonathan Swift noted, "It is useless to attempt to reason a man out of a thing he was never reasoned into."[51] To take a theology at random, the *factual* beliefs of the Church of Jesus Christ of Latter-Day Saints possess about as much intrinsic historical likelihood as those of the Church of Scientology: *Christ resurrected in Mexico? Native Americans as the lost tribes of Israel? Coded golden tablets found buried in Palmyra,*

New York (with accompanying decoder gemstone spectacles)? The faith's true strengths, though, lie outside history or logic; they are *community, moral restraint, mutual reliance.* The more detailed our body of belief, the less likely we are to apply logic, except as a test of consistency with accepted doctrine: hence the sincere worries of contemporary adult Jews, Christians, and Muslims about, respectively, whether deer count as cattle or wild beasts; whether the Host given in the hand, not the mouth, confers grace; and whether a devout man may use a urinal.[52] When we are not explicitly bound by belief, though, we seem perfectly willing to apply other logical systems—thus you should put your trust in God, but still check your coolant level before a long journey. Even skepticism can't be maintained consistently, across the board: Richard Dawkins himself loves singing Christmas carols.[53]

So, if not logic, what is it that makes people fall out of belief or abandon opinions? History suggests that the great objectors to established creeds (Akhenaten, Buddha, Christ, Muhammad, Martin Luther) were driven to dissent, not because they felt the creed made no sense, but because of the hypocrisy of its practitioners. They saw the contrast between the purity of belief and the corruption of religious establishments. Like righteous adolescents, they were enraged at the duplicity of the elders. Even the skeptic physicist Bob Park, author of *Voodoo Science,* scourge of sloppy reasoning and tendentious hypotheses, started out as a perfectly content member of a Methodist youth group; it was only when a pastor answered his innocent questions about Genesis with "you can go to hell as quickly for doubting as you can for stealing" that Park lost faith.[54] The tribe had betrayed his trust.

We began this section with the assumption that hating people for their beliefs was somehow different from hating them for the color of their skin or for their willingness to take low-paying jobs. The truth, though, is that there is no real difference; beliefs are simply part of the compact that makes us *Us*: tribal attributes that strengthen social bonds and help identify outsiders. Perhaps this explains why people are so *sure* about their religious and political beliefs, although they are happy to doubt their certainty on questions like the name of the capital of Venezuela or the number of fluid ounces in a pint. It may also explain why the etiquette books tell us to avoid the topics of politics and religion in mixed company: once off these subjects, people are so much more reasonable.

WHAT TO DO?

In 1939, the poet John Berryman complained that actually to *know* all the news of any given moment would "Curry disorder in the strongest brain, / Immobilize the most resilient will."[55] We are close to that state now. Instant, undigested news puts the Other in our faces all the time, allowing us to extend around the globe those emotions of mistrust and repugnance that we used to reserve for our near neighbors. We watch with horror the glaring eyes and gnashing teeth of fundamentalist rioters half a world away; the same satellites carry them the pouts and posturings of our nymphet entertainers and native-bashing action heroes. The spread of the Internet means that your village idiot is now, unfortunately, in touch with fellow idiots of his kidney all over the world, while the indiscriminateness of the media exalts motivated reasoning to the point where it seems like the real thing: opinion is fact, conviction is proof, "the worst are filled with passionate intensity," and all that.

Most public discourse aims straight for the recipient's limbic system, leaving the prefrontal cortex, that proud badge of human superiority, as unused as a lemur's. We live in an age of conspiracy theories: Bush was behind 9/11 (even the French minister of housing thought this possible), the CIA created AIDS, and the polio vaccine is a Western plot to sterilize Muslims (a canard that has allowed polio to bounce back from near-eradication).[56] By any standard of reasonable thinking, these theories are total nonsense, but the research covered in this chapter makes it shamingly clear that we will swallow *any* nonsense as long as it confirms our preconceptions about the relation of Us to Them. After all, They mix blood with their beer.

How, then, do we get out of this trap? Unfortunately, telling ourselves to apply more reason and less emotion is likely to be just as effective as resolving to eat less and exercise more. Instead, we need to recognize that if the problem is limbic, the answer is probably limbic too.

Mouthwash and deodorant are therefore a good start. Broadening dietary horizons can help: the British generally think well of their Indian immigrants in part because they have come to love Indian food—unaware of the fact that most of their "Indian" curry-houses are actually run by Bangladeshi Muslims. Shared music and dance is a powerful social signal: Edward Hagen at Humboldt University in Berlin suggests that the brain naturally relates synchronized musical experience to the forming of cooperative coalitions.[57] Dancing, he proposes, began as a territorial defense signal, something like the Maori *haka*; so if we all dance together, we all share the same territory. Common laughter initiates and strengthens social bonds, expressing mutual relief at the passing of a potential danger or conflict.[58] Seen in terms of these last two points, Berry Gordy of Motown and Richard Pryor probably did more to improve the relations of black and white Americans than any two politicians. Parenthood, too, tends to trump racism; the young of others, no matter who, are *cute*; we are so deeply programmed to respond warmly to big eyes and little noses that we see all babies as Our babies.[59] The universal ties of worry and pride for our offspring link parents across the deepest cultural gulfs—another argument for fully integrated, secular schooling from the earliest age.

Most of all, we need to seek complexity in the Other: to try to believe that he or she is as distinct, as real, as the 150 people on our Christmas card list. Here, our world-girdling communications technology can help, because it increases the chance for random contact between individuals in drastically different circumstances. One day in November 2000, for example, a woman in Tel Aviv named Natalia Wieseltier got a wrong number: trying to call a local friend, she had instead come through to a Palestinian named Jihad who lived in Gaza. Instead of hanging up, she asked him how he was:

> He said he was very bad, his wife was pregnant, and their town was under a curfew. We ended up talking for about 20 minutes. We weren't making apologies to each other. I wasn't trying to make him feel better. We were just talking as individuals. At the end of the conversation he said he was amazed that Jewish people were able to talk like that. He thought we wanted all Palestinians dead.[60]

Thanks to caller ID, Jihad found Natalia's number on his phone the next day and called her back to say that talking with her had made a real change in his thinking; then his brother called her; then his friends. "I think they all thought there was this weirdo Jew from Tel Aviv who likes talking to Arabs. I started thinking that if we could all talk like that, about the basic everyday stuff, we'd be amazed at how much we had in common." Acting on her thought, Natalia set up "Hello Peace," an anonymous toll-free telephone service that links individuals over the barriers and checkpoints of mutual distrust. Its switchboard has since logged over a million conversations. About what? Does it matter?

The anonymity of the Internet makes us open up far more than we do in physical meetings, telling the world through Facebook things we would not tell our grandmothers. In the end, this may be a good thing for real life too: once we have finished seeking out the few who share our worst instincts, or have tired of ranting at each other in CAPITAL LETTERS, we could take this chance to come to know the Other—for, on the Internet, no one knows what you smell like . . . yet.

CHAPTER VI

Fresh off the Pleistocene Bus

Our times are marked by astonishing diversity and unprecedented speed of change, yet we who inhabit these times are genetically identical to the cave dwellers of seventy thousand years ago. Do our failings in life-adjustment stem from a discrepancy between what we are physically and who we have become culturally? Certainly, many of our characteristic errors—infidelity, obesity, poor parenting, and environmental irresponsibility—track the ways we have diverged from the "normal" as understood by our ancestos. For them, mutual dependence, pursuit of shared goals, and conscientious regulation of social relationships within a small group added up to a way of living that would feel right to most of us, despite the lack of bathrooms, convenience stores, and twenty-four-hour news. This offers a clue to how we can best spot and avoid the errors of modernity: through culture, because culture gives us a way to evolve understanding and attitudes without changing genomes.

The evolution of the human race will not be accomplished in the ten thousand years of tame animals, but in the million years of wild animals, because man is and will always be a wild animal.

—CHARLES GALTON DARWIN (1887–1962),
The Next Million Years, chapter 7

IMAGINE that on one of your trolls through the genealogy sites in pursuit of a wayward great-aunt you came across this exciting invitation: "Meet the Whole Family!" Twenty-five hundred couples, all guaranteed to be your blood relatives, will be gathering in the function suite of a local mid-range hotel. How can you resist? When the day comes, you arrive, register, pick up your name tag, and head in—keeping an eye out for any features familiar from your mirror.

Well . . . they might have mentioned it was fancy dress. Even discounting the gowns, the robes, and the furs, this is a pretty diverse bunch. And those people way at the back—my God! They're *naked*! And look at what they're doing to the buffet!

For this particular reunion extends its invitation, not over space, but time: the twenty-five hundred generations that extend in a direct line back from you, mother by mother or father by father, to the earliest known settlements of fully modern humans more than seventy thousand years ago. A modest convention center can hold them all; if you wanted to restrict the guest list to those ancestors who postdate the invention of agriculture and the domestication of animals, it would include around four hundred couples—about right for a really good wedding. You could fit your direct line over the twenty-five hundred years since the time of Socrates, Buddha, and Confucius into a steak house; that over the eight hundred years since Averroës, Maimonides, and St. Francis into a McDonald's. Every great-great-great since the Declaration of Independence could come to a Fourth of July cookout in your backyard and still leave room for the neighbors.

Time and change are each the other's measure. A life span may be two millennia for a sequoia or a day for a mayfly, but it is still one life. A generation among bacteria can last as little as ten minutes, as opposed to twenty-eight years for humans: it would therefore take a well-fed bacterial colony only fourteen days to go through as many generations as has humanity (another good reason to clear out the refrigerator before you go on vacation). Yet it is only at each new generation that we have the chance to evolve, to shake again the genetic dice-box and test any resulting mutations against the challenges of life.

Compared with the frantic breeding cycles of most living things, ours is pretty sedate—and moreover, we have been working from a small base. It's easy to forget, on this teeming planet, how very few of us there were even recently: a thousand years ago, the population of the whole world was about the same as that of the United States today; at the end of the last Ice Age, about the same as that of metropolitan Detroit.[1] Extrapolation from DNA evidence suggests there may have been as few as two thousand humans in total at the time, seventy thousand years ago, when a bold band of around 150 of them set off from East Africa to populate the rest of the world. We are all descended from that handful, and also, necessarily, from their far later progeny; our ancestry is as inbred as an archduke's. This fact is a simple product of arithmetic: if you consider how many ancestors you must have (two in your parents' generation, four in your grandparents', and so on) and compare this with the past total population of humans, you will find that as recently as thirty generations ago, or less than a thousand years, your 2^{30} potential ancestors would represent more than three times the world's population. So there is at least a statistical certainty that *any* given human of that time is an ancestor of yours. We are not just all one human family; we are also all descendants of William the Conqueror. And if you have blue eyes, then you share with us, the authors, a single ancestor who probably lived near the Black Sea some ten thousand years ago: hello, cousin![2]

Put all these considerations together and it becomes clear that the last seventy thousand years, while certainly eventful as history, have been little more than a tick in terms of physical evolution. Species of wild mammal may have died out in that time, but none have arisen, nor even changed significantly. The naked people at the back of the ballroom may look somewhat smaller and wirier than us, but they are genetically identical—and that includes the genes that determine the structure and function of our brains. Despite the disparities in our outward appearance, we are using the same perceptual tools and instinctive beliefs that they did. Whatever worked for them, we still have—whether it still works or not.

What, other than the statistical constraints of time and change, makes this continuity apparent? Well, there are some basic behavioral assumptions that leap the gap between people from any race or culture on the planet, no matter what their historical experience. Show a Foré tribesman in Papua photographs of a Californian expressing happiness, fear, surprise, or disgust and he will know exactly the emotions this odd, pale baby-face conveys—although the concept of a photograph may be astounding to him.[3] Postures, threatening or welcoming, are universal. We sing and dance at similar occasions for similar reasons: calling on gods, drawing together communities, whipping up fighting spirit, impressing girls. At a deeper level, our phobias reveal a common instinctive underlay beneath our various cultural carpets.[4] Why, when you think about it, should an Irishman fear snakes? Why do Australian children know to be wary of spiders, but not of the equally venomous platypus? Our fears of heights, enclosed spaces, blood, contagion, male strangers—all are general among humans and all seem good evolutionary adaptations to potentially fatal threats. Fortune, after all, favors the bold but punishes the fearless.

A yet subtler evidence of the inner cave dweller is in our attitude to death. Here in the modern West, we tend to keep it well away from our children, not only because we want to avoid traumatizing them but also as a natural result of longer life spans in smaller families: there's less of it around. We don't expect children to understand mortality's conventions, but in one particular case they seem instinctively to know all about it: predator and prey. Children as young as three, independently of any firsthand experience, can accurately describe the sequence of a predator stalking, attacking, killing, and eating prey, without any fanciful or storybook additions, and will tell you that, once killed by the predator, the prey is *dead*, can no longer act for itself, and that this state is permanent and irreversible.[5] They weren't told this on *Sesame Street*—it is part of the human birthright.

Other cues reinforce this sense of a shared mental inheritance. Not only can any baby learn to talk in any of the nearly seven thousand languages of the world, but all mothers speak the same language to their children—a soprano singsong in which the same tones have the same meanings all over the world: that's *good*! What's *that*? No, no, *no*, careful. Oh, poor *baby*. As with swearing, the words may change but the tune is universal.[6]

We also share a sense of what counts as an attractive place. Stephen Kaplan at the University of Michigan (our relative only to the degree that all Kaplans descend from the prophet Aaron) tested the landscape preferences of people in thirty studies from places as widely apart as Australia, Canada, Korea and Egypt.[7] All plumped for natural landscapes over man-made ones and responded particularly strongly to views of green flowery meadows with trees and fresh water—a rather good description of most cultures' idea of heaven. Other researchers have pursued these preferences further and added a general liking for *refuge* and *prospect*: nooks to avoid predators and viewpoints to scout out the surroundings.[8] The grottoes and vistas of the great landscape gardeners are by no means follies, then; they please because they draw on some of our most ancient and sensible instincts. Any caveman would understand, and approve, the work of Frederick Law Olmsted or Capability Brown—and share our unease with the necessary density of a populous world. We envy the millionaire's country estate because we feel expropriated from it.

So if you were to go from the family reunion to visit with your oldest ancestors—those lithe, brown, smiley people—you would find much that feels familiar. They would probably live someplace like Blombos Cave, overlooking the Indian Ocean in South Africa's Cape Province: the oldest site so far excavated that shows evidence of occupation by modern humans.[9] As you settled in, you would likely agree that this is a pretty good spot. The cave is set into a steep hillside, with the view from its mouth commanding all approaches. Shelves and nooks in the walls provide convenient places to sit or to keep important objects. The rocky coastline has abundant food: shellfish, hares, rodents, seals, antelope. Hundreds of bright flowers sparkle in the surrounding shrubland—what the Afrikaners now call *fynbos*. In the near distance, beyond the surf, spout majestic, inaccessible whales.

Life with the Blombos people, like a good camping trip, would be busy but not stressful. The getting of food is the main activity—but, with well-practiced skills, this hardly requires all the hours of daylight. There will be expeditions to seek out the good kind of tool-stone from twenty miles away and afternoons spent in the cave-mouth, flaking it into delicate points. Red ochre is another valuable, useful for decorating things and people, giving them all the ruddy tint of life, as well as for scratching geometrical patterns in, for no apparent reason other than the satisfaction all doodling brings.

The essence of this existence, like the existences of many present-day hunter-gatherer societies, is that it is the *opposite* of nasty, brutish, and short. It has aspects of that ease and certainty we all hope to regain by becoming rich. A small group in a large territory will always have enough to eat, although treats may be an exciting rarity. Intelligent "primitive" people—observant gatherers, resourceful hunters—lead lives of leisure compared with the year-round toil and anxiety of agriculture, or our own wage slavery. They have what we too often lack: the freedom to enjoy the moment.

Say, for instance, you were one of the !Kung people of the Kalahari in Botswana. Your daily assignment would be to find the four thousand or so calories necessary to keep yourself and one unproductive dependant alive. You might have few physical tools available for the job, but do you have expertise. You know the habits of fifty-four species of animal and can identify eighty-five kinds of edible plant, many of them underground tubers that show almost no trace on the surface. You can rank these easily in terms of their likely proximity to permanent water (after all, food without water is useless). You have an accurate sense of how much work—tracking, digging—is necessary for the particular energy value of each food. Thanks to all that knowledge, you need only spend about a quarter of your time actually securing a living, and the rest is free for mending equipment, dancing, singing, and telling tall stories about the lazy man who kept the sun in his armpit before the children threw him up into the sky and illuminated the world. Life, within certain rigid limits, is easy. Even arranging for shelter, that most expensive and troublesome commitment for modern families, is no bother: a leaf-hut takes at most a couple of hours to set up—a tepee, igloo, or yurt not much longer.[10]

Such people have always had time, which allows them to elaborate the traditions, legends, jokes, and speculations that make human life so rich yet leave so few traces in the archaeological record. Time also lets them concentrate on the most important business of all: the regulation and adjustment of relationships. When anthropologist Melvin Konner stayed with the !Kung, he found it impossible to sleep when there was any disagreement within the group; they would talk all night, hut to hut, until the problem was resolved—an encounter group in the middle of the desert.[11] Maintaining relationships is as vital a skill as plant lore or spoor tracking, because without a functioning group, the individual human is terribly vulnerable: unendowed with claws or fangs, slow of foot, dull of sense, easy prey—and there have always been leopards in Eden.

This explains why a visit to Blombos would quickly feel so natural, despite the gulf in technologies, abstract knowledge, and formal language that separates you from your ancestors: the emotional content of relationships, the meanings of the heart, would not have changed. You would need no translator to understand the group's expressions of celebration or grief. You, too, would feel the tension of an approaching quarrel. The children, initially shy, would find your funny faces uproarious and your ability to tie a bow amazing. And over time, you might start to find some, or one, of the tribe members . . . well, rather *attractive.*

LOVE AND MARRIAGE GO TOGETHER LIKE A HORSE AND BICYCLE

Love, love, love. Eternal, indestructible—yet, for some, suddenly gone tomorrow during one snide exchange over the breakfast table. Many waters cannot quench love, neither can the floods drown it, says Solomon, but it might not survive a theatrically pained sigh. It renders the wise foolish and fools witty, yet a sour maxim says it is no more than an exchange of two fantasies and the contact of two skins. Love makes the world go round; it is in the air; it is all we need. Is anything so worth thinking about? Does anything make so little sense?

In theory, love should be simple: the deepest form of partnership, the commitment that makes each other whole. We two form a multitude, and yet are one. The spark of the first glance catalyzes a union of souls, spreading and deepening through a lifetime. The grizzled coot whose arm you hold is still, deep inside, the hot-eyed youth who stood beneath your window; that apple-cheeked grandma still radiates the glow of her bridal blush.

In practice, few even approach this ideal. Teenagers surveyed in schools agree increasingly that a happy marriage and family is the most important thing in life—but also increasingly predict that it won't happen for them.[12] They have reason: the U.S. divorce rate has doubled since 1960. It may have dipped recently, but this is primarily because fewer couples are bothering to get married. Children born outside marriage now account for 40 percent of total births; half of these are to mothers who live alone. On average, an American child born today has a one-in-two chance of being brought up by a single parent before reaching the age of fifteen.[13]

If marriage is becoming rarer, you might think it represents some extra, positive commitment and would therefore be more successful once begun. After all, fewer people suffer the mortifying surprises of the wedding night (or *lack* of surprises: not-so-distant forebears of ours came back from Niagara still virgins through sheer innocence; fortunately for us descendants, a roué uncle soon set the bridegroom straight); more than half of all marriages now begin with a period of cohabitation. Simpler divorce laws *should* mean that those still married remain so because they like it. Yet the number of couples reporting that they are "very happy" in their marriage has continued to slide downward.

Such statistics are alarming. If our automobiles were as unreliable as our relationships, there would be a national scandal. Nor is the comparison of a breakup with a car crash entirely trivial: in both, the deep bruises last far longer than the initial shock. Those to whom either has happened no longer feel entirely safe; they become afraid to take further risks. They have lost control of events and become victims. Their children may no longer trust them, or their own futures, with the old implicit confidence—while the rest of us indulge a morbid fascination in their misfortune, rubbernecking as we cruise by.

The result is bad personally and bad for society. You may think what you like about marriage as a legal form or cultural artifact, but economically it is a machine for generating wealth: "Two are better than one," as Ecclesiastes points out, "because they have a good return for their labor." Two people together accumulate more than two people apart, both through economies of scale (only one tube of toothpaste, one bottle of Tabasco, one house) and through division of work (I'll get the groceries; you pick up the kids). Moreover, the responsibility and pride implied by marriage keep people plugging away at jobs they might otherwise give up: men earn between 10 and 40 percent more after they get married—not, presumably, because they have suddenly become more clever. Divorce, on the other hand, sees women's income drop by 27 percent, while men's rises only 10 percent, so society as a whole is a loser.[14] Single parenthood is the major source of child poverty and income inequality, as well as of direct burdens on the government. One study concludes that each divorce adds an average of thirty thousand dollars in extra welfare support and costs for juvenile delinquency. The 1.4 million American divorces in 2002 therefore cost taxpayers more than forty billion dollars, about a quarter of the federal deficit for that year.[15] The environmental impact can be as great as the financial: divorced households use between 41 and 62 percent more resources per capita than when they had been married households—in the United States in 2005, that translated into more than 38 million extra rooms, 73 billion kilowatt-hours of electricity, and 627 billion gallons of water.[16] For what is supposed to be the most natural urge of all, love is causing us a lot of trouble.

The global question is the same as the one so many ask as they lie awake at three in the morning: *why are we doing this*? What is so wrong with our romantic vision that it makes us walk face-first into disaster? And why, if we are still using the brains we had as cave people, do we not have some simpler, pre-civilized way of arranging our love lives?

These are questions more of heart than head, so their answers may not be those of statistics, but of literature:

> Before her marriage, she had thought she had achieved love; but as the happiness this love should have brought did not appear, she thought she must have been mistaken. Thus Emma tried to grasp what exactly was meant *in life* by those words—"joy," "passion," "rapture"—that had seemed so beautiful to her in books.[17]

In Emma Bovary, Gustave Flaubert created one of the most troubling heroines in literature. Even if you haven't read the book, you know the story: oppressed by the stupendous dullness of provincial life and a boring if inoffensive husband, Mme Bovary seeks the romantic fulfillment that her heart and her reading insist is possible by falling into intense affairs with a student and a rich landowner—but she gains from them only betrayal, indifference, exploitation, and despair. Even her "romantic" choice of arsenic for her suicide, taken in the hope of expiring in wan forgiveness of the weak and venal men who had ruined her, is a false hope: she dies only after a weeklong agony of vomiting.

The novel was not intended as some primly moralistic tale. Indeed, Flaubert insisted that he *himself* was Emma Bovary: his exposure of her romantic passion as an addiction to empty sentiment was really a failed attempt at exorcising his own romanticism, his unrequited devotion to every detail of an indifferent and random life. When Charles, Emma's fiancé, notices how "the way she brought her hand to her hair, the sight of her straw hat hanging from a window-lock, and many other things in which it had never occurred to him to look for pleasure—such now formed the steady current of his happiness," we see captured the novelist's and the lover's logically false conviction that here and now—and *she*—are different from any other example life has ever offered.

Unfortunately for all romantics, love's root cause is probably parasites. Asexual reproduction by budding may be all very well for bacteria and plants, with their millions of progeny. But we larger, longer lived, less fecund creatures need to keep our potentially devastating parasites guessing by shuffling our genomes every generation through sex. Epidemics are the punishment for purity, so—despite the objections of racists—we are probably very wise to find that the thrill of the stranger includes attraction as well as fear. This was neatly confirmed in the famous Swiss T-shirt test of 1995: a variety of men were asked to sleep in the same T-shirt for three successive nights. A range of women were then given the shirts to smell and asked, on that evidence alone, which of the men they found attractive. The result was remarkable: each woman chose the man whose immune system (as measured by his major histocompatibility complexes) was most different from her own.[18] Were this to be her mate, she would be ensuring that their offspring had the benefits of as wide an immune response as possible. The implications for the aftershave industry are obvious and worrying.

Sex is a strategy, though it's never as straightforward as the "scoring" or "man-catching" techniques that feature in the lads' and glamour magazines. It is a strategy to secure, by a fundamental compromise, a future for as many of our genes as possible. Since we cannot yet clone ourselves, we must come to an arrangement with the other gender—which means balancing interests that are not necessarily identical. A healthy human male could, if he made best use of every sperm, father some twelve million children every hour. A woman has a total of roughly four hundred eggs, but a likely maximum progeny of around twenty children in a lifetime; most present-day hunter-gatherers have between three and eight. This essential discrepancy in potential offspring shapes the prototype of sexuality: males intent on their moment of pleasure, repeated as widely as possible; females all too aware of the months of awkwardness, hours of pain, and years of care this moment implies.

Roving men, harried women: the archetype is lived out everywhere populations are large and fluid and chances for long-term success are few, in slums and shantytowns all over the world. But this is not how affairs went during the first 99 percent of human history: over all those thousands of years, we lived in small family groups, where the survival of a man's genes depended on much more than simply finding someone to carry them for him. Such simpler lives depend on stronger, deeper relationships—ones that may no longer be matters of life and death, but still feel that way.

A human child, as every cash-strapped parent knows, is an investment. From its first morning, a baby deprives its mother of one arm, one hip, and half her attention. In a world where food must be found every day, this puts both of them at risk: fruits from the thorny bush, shellfish from the deep pools become inaccessible. What's needed is a coinvestor—someone equally interested in this particular baby's chance to grow up through years of childhood to breed in turn: a husband. From the man's point of view, it may be all very well to mate promiscuously, but how will you know which of the children in your group are actually yours, to be fed and trained into successful adults? Male langur monkeys, like furry Herods, try to kill all the infants when they take over a mating group, but such a bloody solution would not work well for humans, because infancy lasts so long. What's needed is a coinvestor who can offer sole rights to paternity and special care for those few guaranteed offspring: a wife.

This all sounds pretty heartless, the stuff of medieval dynastic alliances or grim tribal practice, where reproductive capacity is a commodity to be traded or hoarded through arranged marriages and polygamy. We modern Westerners feel, with Emma Bovary, that romantic love should take precedence—and secretly hope that we can have it all: perfect children at just the right time, support, freedom, understanding, and the *grand amour*. We describe ourselves in the personal ads as potential lovers, not potential parents: we are *passionate*, *frank*, *feisty*, *gorgeous*, *fun-loving*—all with, of course, a GSOH. Nobody gets dates by saying "reliable reproducer seeks same to merge genomes."

Scientists, though, can see through the lace and feathers of courtship to the needy body within: many of the preferences we attribute to love's notorious capriciousness are actually straightforward assessments of reproductive fitness. Men prefer the faces, voices, and smell of women who are in mid-cycle, at peak fertility, even if they have no idea why.[19] They respond optimally to a female waist-to-hip ratio of 2:3—"childbearing" hips—although their preferences for overall plumpness tend to depend on circumstances.[20] Men where food is less plentiful prefer heftier women, and even the same man, when hungry, seeks a more curvy mate than when full. So if you are still losing those last five pounds, don't let him take you to dinner; an extended late-afternoon walk will better whet his appetite for you.[21]

Women, as befits the partner with more to lose, have more complex requirements. They like tall men and place great emphasis on symmetrical features—a good indicator of general health.[22] Indeed, although women say they associate a deep voice with height, hairiness, robustness, and other desirably masculine characteristics, thus finding it attractive, it actually turns out to be a better predictor of symmetry than of high testosterone—which is why, you men, even if you don't look as macho as you sounded on the phone, you may still have a fighting—or, rather, loving—chance.[23] The much-vaunted trio of eyes, hands, and buttocks as a basis for first assessment also has sound selective reasons behind it. Eyes: unless you show the involuntary contraction of the *orbicularis oculi* muscle (the so-called *Duchenne smile*), no amount of toothy grinning will convince a woman that you are a nice guy—and genuine kindness is essential in someone who will be having to supply food to mother and children over a long time. Hands: these, as we know, are a good indicator of testosterone levels, but they also show, through bone growth, how well nourished this potential provider has been. Strong but long, dexterous fingers are therefore a good sign. Buttocks: these, as much as symbolic thought, distinguish us from the other great apes—and for an interesting reason. Daniel Lieberman of Harvard has suggested that a change in environment from forest to savanna put our early hominid ancestors onto their two feet: prey stopped being something you could ambush in the trees and started being something you ran after, mile upon mile, until it eventually succumbed to exhaustion.[24] We had our uniquely human buttocks even before we had our unique brains, and a lean, muscular backside is, like a single-minded ambitious intensity, a promise that this man will go great lengths to bring home the antelope.

Human males are a variant of females, produced by the action of that most volatile of hormones, testosterone. As an anabolic steroid, testosterone generates those specifically male characteristics that are useful in a life of divided labor: bulk, upper-body strength, daring, aggression—and, we could even say, a linear, persistent habit of thought suitable for pursuing prey to the exclusion of all distractions (especially housework). Too much testosterone, though, acts as a dangerous solvent of relationships, generating violent rages, infidelity, and indifference to children. Besides, it's bad for the health: castrati live as long as women, he-men do not.

Women can spot a man's overall testosterone level simply by looking at a photograph of his face.[25] The strong jaw, prominent nose and eyebrow ridge of a "dominant face" will predict success in a predominantly male environment like the military or a professional sports team. But women are not so easily swayed: they gauge these faces as cold, untrusting, and bad at fathering. In the same study that demonstrated women's uncanny hormone-reading ability, the men with lower testosterone levels were far more likely to report themselves as being fond of children; and the women chose these men as preferable for long-term relationships.

"Long-term" can have many meanings, though. In the long term, it is obviously good to have a kind and devoted husband. But is it good in the yet longer term to bind your own genes' futures solely to such a domesticated lineage—to put all your eggs in one behavioral basket? Hominid evolution reflects great changes in circumstances; we have not survived so long through cozy conservatism. It may not, therefore, be a good idea to breed sons who are all dependable, pipe-smoking, slipper wearers; some may need to be risk-takers, willing to seek out new feeding grounds, able to face down rivals for supplies, even if they make bad fathers. The arts of war are not those of peace; stressful times call, briefly but insistently, for the powers of testosterone. An unconscious awareness of this discrepancy may explain why the same women who chose low-testosterone men for long-term relationships preferred the faces of high-testosterone men when contemplating a brief fling.

If sexual reproduction is, in essence, a way of spreading one's genetic bets, it can be argued that a little well-concealed sexual infidelity is a reasonable if cynical strategy.[26] Among the Canela, Ache, Bari, and Yanomamö people of the Amazon, this idea is enshrined in the ingenious theory that children are formed like pearls, through successive layers of male insemination—so not only is a woman justified in sleeping with a range of desirable men; she can also be assured that they will *all* consider themselves fathers of any subsequent child.[27]

Emma Bovary, therefore, followed a reasonable path when, having secured the boring but devoted Charles, she pursued risky affairs with the high-testosterone alternatives, the passionate Léon and the philandering Rodolfe. Her only mistake was in thinking them worthy of trust or capable of real love.

Looks and dependability, buttocks and testosterone are not the only elements in the calculation of female mating strategies, though—they are not even the most important. Remember that children are an investment, so a father must not only be *willing* to invest (thanks to his kindly temperament); he must be *able* to invest, thanks to his proven capacity to provide resources. Material success remains the most important recommendation: women rate it twice as highly in selecting a mate as do men, in studies made over the past seventy years and in thirty-seven cultures around the world, from Finland to Brazil, South Africa to Iran.[28] In personal ads, women seeking men mention financial prospects eleven times as often as do men seeking women; U.S. college women would "not be happy" with a man who made less than the highest 30 percent of national earnings.[29] Being a billionaire, on the other hand, gains forgiveness for many other shortcomings. From *Pride and Prejudice* to *Bridget Jones*, the hero's fortune will redeem his faults.

It takes longer training to become a good hunter than it does to become a brain surgeon. Starting at age five with tiny bows and quill arrows, Amazonian, Papuan, and Kalahari boys do not reach their peak of hunting skill until their thirties, more than making up for their declining physical powers through expertise, lore, and guile. It makes sense, then, to match the peak family demand for meat with the father's peak ability to provide it. And, in fact, women generally choose men who are older than they are to make this happen—by four or five years in many hunter-gatherer societies, but up to twenty-six among the gerontocratic Tiwi people of northern Australia.[30]

What about men's preferences? Youth and potential fecundity matter a great deal, as any *Playboy* reader could tell you. Beauty is important—and is not at all in the eye of the beholder: people from all the American races and from cultures as distinct as Chinese and English or Ache Indian and Russian will agree on which face out of a group from *another* race or culture is the most beautiful.[31] Surprisingly, the attractive is often the average: people find a composite photograph more attractive than the individual faces from which it was created.[32] We shun the asymmetrical and the extreme, except in the rare examples where excess succeeds; but if some Frankenstein of a plastic surgeon were to create a face with Scarlett Johansson's lips,

Gong Li's eyes, Aishwarya Rai's nose, and Renée Zellweger's cheekbones, most men would probably flee the result as a monster.

In behavioral terms, men, unsure of paternity, traditionally place great importance on female chastity, often using the token of virginity at marriage as an indicator of future faithfulness. Even in more free-minded societies, where women are in control of their economic and sexual destinies, men are irrationally sensitive to a potential mate's faithfulness: frenzies of jealous passion are still known among the tolerant Dutch and liberal Swedes. For themselves, though, men require less sexual exclusiveness. Indeed, in what's called the "Coolidge effect," male sexual performance increases at the prospect of new mates.[33] As if in resigned expectation of this, women tend to react more strongly to emotional than to sexual betrayal: if he really loves me, not his mistress, life can go on; but if his heart has also left the bed, I will be dreaming of revenge.

There is one further aspect of appearance that men and women value equally highly: clean, thick, lustrous hair. To judge by the over five billion dollars spent annually on hair products, this is no mere passing impression, but a fundamental aspect of attractiveness—or why would so many puritan cultures make head-covering obligatory? Good hair is, certainly, an indicator of health and freedom from parasites. In men, it shows a desirable intermediate level of testosterone, between the thin crop of the eunuch and the bald dome of the excessively dosed. The primatologist Alison Jolly, however, speculates that good hair, nicely styled, may actually be a social signal.[34] Among the other primates, grooming is a sign of status: the more important you are, the more time others spend ruffling through your fur, leaving you tick-free, sleek, and . . . well, *well-groomed.* Furless humans don't easily display the results of such attentiveness (though we try to make up for it in body decoration and ornaments); instead, a well-combed or cornrowed or otherwise "done" head of hair demonstrates the support you command among the group. Someone, on the other hand, who has to cut his own greasy fringe with nail-scissors and Scotch tape will not have much to offer a potential mate.

If you were to post your personal ad on the wall of Blombos Cave, then, you would be wise not to spend much effort describing your passion for long walks, jazz, the Republican Party, or Fassbinder movies. The ideal male would say something like this: "Successful hunter, well-regarded, with several chunks of ochre. Tall, broad-shouldered, attractively shaggy, nice buns. Fond of children but don't have any. Ready for commitment. Age: about five years older than you." The female would reply: "Youthful, symmetrical, flowing-locked daughter of skilled gatherer. Perfect waist-to-hip ratio. Faithful, observant, cautious, tolerant of peccadilloes. Fond of children but don't have any."

We can see many of these ancient assumptions at work, if not in our own grand passions, at least in the current of relationships that swirl around us. When women want to denigrate a rival to a man, they tend to disparage her physical appearance or her sexual faithfulness ("what, that *fat slut*?"), while men attack status and ability to provide ("what, that *wimpy loser*?").[35] High-status males, as they age, recycle their prestige through ever younger wives; high-status women may have toy boys, but these are not considered a breeding relationship—yet one more reason why powerful women have few children. The unfairness of this social prejudice is obvious, but that doesn't make it easy to overcome. No Hollywood producer is likely to make a movie in which a work-obsessed female highflier sweeps into a downtown diner to carry off the hunky if ditzy short-order cook in her limo to share a life of luxury—because every audience will instinctively think, "there's something wrong with this picture."

Thus far the enemies of romance. If you have any soul at all, you will probably be grinding your teeth by this point: not for nothing has evolutionary psychology been called the "moral equivalent of fast food." Reducing all of love's exaltation—that sense of becoming a better person for the best of reasons—to some gonadal imperative can be a depressing and shabby-feeling enterprise. Was Dante lovestruck by

Beatrice's hip-circumference? Unlikely: she was only eight, after all, when first he saw her. The longing and delight of love must surely wipe out such calculation; Cupid laughs as hard at biologists as he does at locks and anxious parents.

The real point, though, is this: we play for ourselves in love's game, but our ancestors gave us the cards. The heart-wrenching power of our feelings reflects the life-or-death concerns of a simpler, more isolated, more dangerous world. Our current troubles arise in part because each is no longer as essential to the other as we used to be. We know that, physically and economically, we *could* survive and breed as individual units—which makes the challenge of staying together no longer a matter of keeping our necessary side of the parental bargain, but of living up forever to the image formed by that first blazing romantic attraction.

It is an attraction that our ancestors would have recognized. The ancient stanzas of the Song of Songs seem the transcript of any longing heart today. Sappho's jealousy is just as raw as a teenager's diary entry (though somewhat better expressed). Humans have apparently always felt romantic love. Which brings up the question, what is it *for*?

It may be a residue of our long-ago isolation, when our contact with people outside the circle of incest taboo was rare and brief. Fewer choices make us more interested in the remaining ones: for example, it is not just in songs that everyone looks better in an emptier bar or at closing time; it's a tested fact.[36] People in such isolated spots as Meeteetse, Wyoming, or Coober Pedy, Australia, still somehow find someone in the neighborhood who feels about right for them. On the other hand, many lonely people now living in huge cities know what happens when one has too *much* choice; the good man or woman right here may not be the Perfect One, who's just about to round the corner—and every airbrushed commercial image promises that perfect people do indeed exist out there, somewhere. As we wait to give ourselves to the beau or belle ideal, we fail to commit to any of the real, perfectly nice people who cross our paths. This is a poor strategy. A game-theory simulation of courting behavior concluded that there is little advantage in actually dating more than about 10 percent of locally available potential mates. So holding out for Mr. Darcy is as bad a plan as drawing to an inside straight.

There is a reason why *Pride and Prejudice*, like all romantic fiction, ends with the wedding—in real life, the love story has hardly begun. The collection of Farm Service Administration photographs at the Library of Congress contains haunting images of poor rural American couples from the Depression years. Not one is romantic: the men are grimy, lined, and worn; the women look like calico sacks tied around some arbitrary middle with string. They have, though, an epic quality all their own, one that bears scrutiny and speculation. As the Kansas farm woman looks at her husband over the Sunday pies at the church social, what does she see? The shocking whiteness of his forehead as he briefly removes his hat for grace? The twist in his forearm from the badly set childhood fracture? The closed, unselfconscious expression of someone whose little scrap of mirror reflects only his chin and razor? Or the man whose troubles and hopes are also hers? The man with a name and character among others, from the pastor down to the boys at the feed store? In all the glances between wives and husbands, there seems to be something beyond attraction: there is *history*. A shared story of effort; a mutual appreciation of skills, resources, and endurance.

The old ballads where man and woman exchange their tasks with disastrous results were not sung to assign an upper hand to either but to acknowledge how far survival depends on *each* working for *both*. In every society we celebrate this partnership by praising the bride and groom together: "Will his love be like his rum, intoxicating all night long? Will she be the perfect wife and make him work hard all his life?" Attraction and romantic love are understood in even the most remote societies as the inducement, not the reward, for life's partnership. The shared joys and anxieties, the merged memories, the slow-growing habits of accommodation and appreciation—in a successful union, all these come to take due precedence over the restless self, with its drives and dissatisfactions. Now, though, that modern work has moved away from the specialized and demanding roles of gathering and hunting, hoeing and milking, we too easily forget to admire and compliment each other (even if it's only that implicit compliment, dependence), because we each could, at a pinch, do it all. We have a machine, for God's sake—why can't *he* wash his shirts? Would it be too hard for her to check the oil once in a while? Scrub your own damn back! The corrosion of selfishness is never far away: in the absence of absorbing shared challenges, the machinery of partnership becomes stuck at the point of personal romantic fulfillment. This may explain why, at least in some studies, arranged marriage apparently leads to greater long-term contentment than a love match does.[37]

If we want to avoid this trap, we must keep reminding ourselves that Emma Bovary's marriage is not the only literary archetype: there are other, older, more encouraging stories. When Odysseus at last came home to Penelope, any ancient listener would have thrilled at the essential rightness of it. True, he had spent the previous ten years sleeping with a nymph, but during the days at least he had sat on an empty shore gazing longingly toward Ithaca, mingling his tears with the imprisoning sea. She, peerless among women, never forgot herself, though handsome suitors, in all the arrogance of their honed and shining bodies, lounged in her lord's hall. They were ideal adulterers—she, though, knew better. The triumphant reunion of wife and husband was the victory of memory over temptation; their shared trust and experience was the olive tree that bound their bed to the earth.

GETTING STUFFED

> Having filled your quails with the salpicon of truffles and foie gras, then roasted and glazed them briefly with their cooking juices in a hot oven and popped them into the pastry cases, add the sauce you made previously from the shredded lean ham, thyme, bayleaf, game

> consommé, and sauce espagnole—all simmered together, reduced, and strained. Garnish before serving with poached chicken quenelles and cockscombs, lightly sautéed sliced mushrooms and shredded truffles, adding a touch of Madeira to finish.[38]

Hungry yet? Food is the pleasure in which we all come closest to the response of the addict, all limbic system and no frontal cortex, where the howl of desire drowns out the voice of reason. Clear as the theoretical distinction between gourmet and gourmand may be, few of us can exercise any self-control in the face of a platter of pumpkin tortelloni. As Coleridge said, no man can have a pure mind who refuses apple dumplings.

A little more refusal might, of course, be a good idea: the last twenty-five years have put a horrifying strain on the nation's scales. We are getting fatter, faster, than ever before. Taking obesity to be defined, generously, by a body mass index of 30 or higher, only ten states reported more than 10 percent obese people in 1990; none reached 15 percent. By 1998, no state had *less* than 10 percent, although none reached 25 percent. By 2006, however, only four states—Colorado, Connecticut, Hawaii, and Massachusetts—had obesity rates under 20 percent; all the rest were over 25 percent and two—Mississippi and West Virginia—were over 30 percent.[39] And this is not just about being big-boned, zaftig, roly-poly, pleasantly plump; no, this is a situation where more than a quarter of American adults could be heading for major health problems, caused solely by extra body fat. As for just being overweight, that category includes 66 percent of U.S. adults. Still hungry?

What's happening to us? To begin with, we are indeed eating more: average daily caloric intake rose between 1985 and 2000 by 12 percent, or 300 calories, with no corresponding increase in physical activity.[40] Assuming that the earlier intake was sufficient simply to maintain a constant national weight, the second represents enough to gain an extra pound of fat per person every ten days or so, or more than thirty-five pounds in a year. And it's so easy to slip in these extra calories: just one large caffe latte with whole milk from a well-known chain contains 290 calories, without even adding sugar or the blueberry muffin that goes with it so well.[41] You would need to eat six *pounds* of lettuce and cucumber salad to reach the same total. Skipping lunch in favor of a snack may actually be a dietary step backward.

Obesity is bad for health in its increased risks of high blood pressure, diabetes, heart disease, and certain kinds of cancer. It is bad for self-esteem; it is even worse for the esteem we should expect from others. Not just poorly brought-up children sneer and roll their eyes at the overweight: even medical professionals, when surveyed, reveal that they tend to regard obese people as being lazy, stupid, weak, ugly, and bad.[42] Shakespeare's Caesar may have wanted men about him who were fat, but he was in the minority; and the contempt people generally express for obesity may add to the perceptual distortions that lead to its equal and opposite nightmare, anorexia.

All this is familiar—too dispiritingly familiar. With our manifest talents and knowledge, we can't keep our bodies in the shape we'd like without obsessive, tedious effort. Our self-reforming regimes combine a constant search for the new with a sneaking sense of hopelessness: of those people who actually complete an effective diet and lose 10 percent of their body weight, two thirds gain it again within a year and almost all are back where they started within five.[43] The reason the thirty-billion-dollar diet industry is still in business is that it has a lot of repeat customers.

Humans are unique in finding food such an emotionally complex issue. One need feel no sympathy for the wandering whale gulping its daily tonnage of cold raw krill—because food, to whales, is neither a source of anxiety nor of pleasure; it's fuel. Primeval humans, though, had to seek and select their food intelligently from an unforthcoming and potentially fatal environment: a perpetual treasure hunt whose prize was survival. The peasant paradise where pigs run around ready-roasted and candied fruit drops from the trees has never been our fate; we have always had to motivate ourselves to the task of finding nourishment—

choosing the nutritious but not the toxic tuber, deciding which animal is worth the chasing. This may not be a full-time occupation, but it is an absorbing and challenging one: you have to *think* about it, not just graze your way through another dreamy day. Such an obsessive pursuit needs a powerful spur: hunger.

How then did our ancestors manage to stay slim and healthy when we cannot? Simple: they were hungry most of the time. We think of hunger as an undesirable and unnatural state to be resolved as quickly as possible, but in fact hunger is so strong an impulse because it needs to activate a sophisticated and sustained response: the heightened, stealthy vigilance of the hunter, the keen senses and memory of the plant-finder. Hungry animals live longer and age more slowly; the Okinawans, whose culture includes the concept of *hara hachi bu*, or eating until only about four-fifths full, boast quadruple the number of centenarians per million people as do Americans.[44] The hormone released by hunger, *ghrelin*, sharpens perception, boosts learning capacity, and stimulates locomotor activity in preparation for the quest after nourishment (which is assumed to involve something more strenuous than looking in the fridge).[45] Your personal trainer is therefore right in telling you that the proper response to feeling hungry is to exercise—the problem is it's hard to convince yourself that running to nowhere on a treadmill shares any of the thrill of the chase. Perhaps a video of a fleeing but weakening antelope would make it seem more worthwhile.

Maintaining a healthy diet in the face of abundant food requires self-control—something that was never an issue in ancestral life. If your tribe managed to down a giraffe or came across a marula tree in full alcoholic fruit, the correct response would be, "Whoopee! Cram your belly full—such an opportunity may not occur again!" The hormone that induces satiety, *leptin*, tends not to rise immediately in response to local overfeeding: our metabolism is apparently tuned for a low-calorie life with occasional blowouts.[46] In practice, certainly, the feasts of hunter-gatherer societies are titanic though infrequent; and as if in anticipation of this, some populations, such as Native Americans, were said to carry a "thrifty gene": a set of adaptations that prompt the body to store as fat the extra nourishment from these feasts to carry it through the famines to come.[47]

Now, though, we can find all the makings of a top-notch potlatch at any convenience store. No wonder, then, that widespread obesity is often the immediate result of the arrival of "civilization" in traditional hunting societies. Probably the worst scourge of many an Indian reservation—producing a death rate from diabetes *four times* the U.S. national average—is not alcohol, but potato chips and soda pop.[48] Without famine to balance feast, the thrifty gene becomes miserly.

We condemn fat, salt, sugar, and red meat as "bad" foods—but don't really believe it, because they taste so good. What actually sets them apart is not their sinfulness but the fact that in primitive life they are both necessary and extremely rare, and thus to be treasured. Meat and sugar provide concentrated nourishment; salt is essential to nerve function. Fat is a valuable source of energy and important for both breeding fitness and a healthy brain. Primates fed a low-fat diet become more aggressive and less sociable (which sounds familiar to any dieter).[49] Sharing fat is an important part of maintaining group harmony in hunter-gatherer societies from Botswana to the Arctic. It's an essential part of a courteous greeting, as we can see from Captain George Lyon's account of the first contact between Inuit and Englishman: "I washed my face and hands, making [his Inuit guest, Ay-ô-kitt] do the same; during the operation, I saw him cast many longing looks at the tempting piece of yellow soap which we were using, until at length his repeated Ay-yââ's of admiration determined me on making him happy, and he devoured it with delight."[50]

"I love you like salt," said the princess in the fairy tale; and love can also be sweet as honey. Perhaps unfortunately for us, we can today find all these things our ancestors most desired wrapped up in one compelling package: the french fry. First sold on the streets of Antwerp as the cheapest possible imitation of a complete hot meal, it offers a combination of deliciously caramelized sugars, perfectly presented fat content, saltiness, and a touch of meaty umami flavor from the beef tallow in which it was traditionally fried (and which was McDonald's essential secret ingredient—until people in India found out about it). It epitomizes our slavery to primeval wants.

Such powerful desires cannot remain purely a matter of instinct for an intelligent animal: just as love is much more than simply sex writ calligraphically, our feelings about food go beyond mere filling and emptying. As one dieter said: "I was missing what probably had been my best friend over the past few years. Food as friend. Food as tranquilizer. Food to assuage loneliness. Food that enabled me to eat myself into the oblivion of sleep. Food the comforter."[51] We say that eating too much is a failure of self-regulation, but it may actually be a successful attempt to regulate something else: emotion. Hungry people will eat more if they have been told something depressing; stress and worry lead us inevitably toward the kitchen. This may be a useful response for a primitive group (get food in the camp and we'll all feel better; share some eland fat with the bereaved and you'll strengthen your social ties), but it's not good in a world with such constant and varied sources of unease. Today's news offers distinct alarms from all five continents: bad things that might conceivably harm those we love and over which we have no direct power. No wonder Americans reported higher levels of anxiety in the 1990s—yes, even before the war on terror—than they had in the 1950s, the self-described "age of anxiety." Starting in the 1980s, normal children were experiencing more stress than adult psychiatric patients in the 1950s.[52] Every year, about forty million Americans suffer an "anxiety disorder"; it's not surprising so many succumb to the siren song of that tub of butter pecan.

This sounds like self-medication, and the natural response is to assume that there is some medical solution to what looks like a public-health problem: a missing ingredient to rebalance our dietary humors. There are people who willingly substitute for their fat intake the indigestible Olestra, apparently unfazed by the possibility of having to wear adult diapers. Others forsake sugar for artificial sweeteners, unaware that a recent study shows that rats, at least, eat *more* calories and put on more fat when eating food sweetened with saccharin than with sugar.[53] Yet others attempt to go back in time, hoping that by reproducing a Pleistocene diet they can recover the figure of the svelte and elegant caveman—although what they generally gain is the caveman's raging hunger for something fatty or sweet.

All these shifts and stratagems, though, ignore the French. Infuriatingly, Frenchmen eat three times as much butter, three times as much pork, and nearly twice as much cheese as Americans—yet their incidence of coronary heart disease is less than half that of their U.S. counterparts.[54] They drink without shame, they dote on rich sauces and sugary pastries, and yet they can still fit into their wedding clothes at an age when most Americans are discovering the convenience of elasticated waistbands. What is their secret? When *60 Minutes* suggested in 1990 that it might be the chemical resveratrol in their red wine, the world reached eagerly for its corkscrew—but Scott and Ashley have still not shrunk to the size of Yannick and Marilène. To properly understand how French people stay thin, you need to have dinner with French people.

It takes two hours. The courses are many but small. Each is an object of judgment and comment, because each deserves attention. The quail recipe we gave at the top of this section contains, ounce for ounce, fewer calories than a cheesesteak sub, but the rich complexity of its flavors renders it, bite for bite, much more satisfying. By making the consumption of food as much a culture as its preparation, the French (like, to be fair, the Italians, the Spanish, and other "slow food" devotees) increase their pleasure per calorie and give their leptin ample time to kick in and suppress appetite before that fatal moment when the dessert menu arrives. Their ceremonial approach to eating is a hearty acceptance, rather than a guilty denial, of food's unreasonable fascination—and thus puts Parisians much closer, for all their sophistication, to the leisurely, obsessive habits of our ancestors than are we, whose food ritual often is putting a tray into a box and waiting thirty seconds for it to go "ping." The real slimming effect, therefore, stems not from *what* you eat, but *how*: eat with other people, with conversation, with utensils, with manners, and with appreciation—not at a desk, not on the street, not in front of an open fridge at two in the morning.[55]

Wanting the rare, valuing it in imagination, and celebrating its arrival are the three taproots of culture. It is culture that makes going on a fishing trip with your friends so much more appealing than just visiting the seafood counter. Ceremonies and traditions, manners and recipes, are there to sharpen and ennoble life, engendering both a hopeful anticipation of the future and a keen pleasure in the present moment. This is the real gift from our ancestors—and we lose it at the peril of more than our waistlines.

SOMEDAY, MY SON . . .

"Don't tell your child to be an engineer or be this or that, because we have no clue where future jobs will be," proclaims one. "The world is developing so rapidly, whichever job you recommend now will be out-of-date by the time they are out of university," adds another. And who, exactly, are these blear-eyed seers, these self-doubting prophets? The world's political and business leaders, that's who—panelists at the 2008 World Economic Forum in Davos, offering their expert opinions in response to the question, "What job should my child take in a globalizing economy?"[56] By the end of the session the best advice this highly qualified group could come up with was to learn a foreign language, gain some people skills, study other cultures. Gee, thanks, Dad.

Almost three thousand miles south of Davos, the Mbuti Pygmies of the Ituri Forest are setting up their huts. As always when the group plans to stay in one place for a while, the young people create a parallel camp, the *bopi*, within sight of, but off-limits to, their parents. Here children imitate in play the things they have observed adults do: hunting animals, preparing food, singing and dancing, even making love. The boys and girls are organized into age-classes, but their social contacts are general and fluid. They call all women in the parental camp "mother"; they follow the hunters or tag along with the groups seeking out nuts and porcupine burrows. The older children pass on the lore they have picked up from the grown-ups. By the age of five or six, Mbuti children know, at least in the abstract, all they need to know as adults; from then on, they need only practice and improve.

These Pygmy children are doing pretty much what the heavy hitters of Davos prescribed: gaining social skills, assembling the elements of culture, getting a sense of how things work and how people differ. The songs and rhymes they absorb almost incidentally along the way prepare them for the tasks of adult life and the maintenance of precious traditions. Growing up means only that the play becomes progressively more earnest and more productive. Isn't that how it should be?

Perhaps—but if you are a parent, you will know this isn't how it *is*, at least for us. Statistical and anecdotal evidence conspire to frighten us over the deterioration of childhood. When a third of preschool children in one survey said their chosen career was "celebrity"; when we are spending more on ADHD medication for children than we are on antibiotics; when the favorite drugs of abuse among teenagers are anxiety-reducers and painkillers; when twelve million American children and adolescents are diagnosed annually with some form of mental illness; when suicide becomes the third-highest cause of death among fifteen- to twenty-four-year-olds—something is wrong.[57] Our children are so *anxious*—but they lack motivation. They are at risk from predators, bullies, and gun-wielding crazies—but they never seem to go out to play anymore. They are worryingly sophisticated—but unless we beef up their résumés with Sunday soccer leagues and flute lessons, they'll never get into college. They mock our unfamiliarity with the latest developments in technology and entertainment, yet they look to us to arrange their futures—when even the billionaire oracles at Davos have *no clue* what their futures might be. Psychologists' reports, repeat prescriptions, Net Nanny settings, parent-coach preseason briefings, personal essay "consultants"—ah, how sharp are the joys and how sweet the remembrance, when days of my childhood to mind I recall.

Where the ancestral pattern of child rearing differs most from our own is in its close resemblance to the social habits of the other primates. Sarah Blaffer Hrdy, doyenne of social primatologists, has long insisted that the parent-child relation is only part of an upbringing: we are evolved to be *cooperative breeders*, sharing the task of raising our offspring through an extended network of relatives and allies ("allomothers," as she playfully calls them).[58] This isn't simply so we can have some time to put our feet up; it is essential to the health of mothers, the development of children, and the well-being of the group. Each relies on this distributed childhood to create and reinforce the complex relations of character, memory, and emotion that we call *meaning*—the qualities that give human life its grain and flow.

You'll notice that many of the mental illnesses that plague modern Western children are problems of poor socialization: overactivity, withdrawal, obsession, narcissism, rage, despair. When isolated from their allomothers, they lack the thousand little boosts and restraints, delivered by people of varying degrees of closeness, that shape the social self. Socialization is the essence of a primate infancy: just as lion cubs learn to pounce and bite, we learn to assert and submit, negotiate, empathize, and chaff. Infants soon find out when the top male is going to be indulgent and when he is just as fearsome as he looks; they gain a feel for the difference between purposeful work and rumpus time, when it's fine to be boisterous; they see what true warmth in parenthood is (something they can pass on to their own children); and they come to understand the meaning of that much-abused term *respect*: incorporating others' wishes into your own.[59] Without such detailed and intensive training, the group could not survive, because it would soon cease to be a group. The nourishment humans require to grow up properly is not just a matter of enough milk and green vegetables; it's taking in and finding a place in a shared, living culture.

Cultures and situations naturally differ greatly across the world, but the brain's plasticity easily accommodates their variety. We humans begin almost literally open-minded: our brains start out with the potential to form a wide variety of possible neural networks. Only in adolescence do we start to slough off unneeded neurons and insulate our most-used cortical pathways with myelin. This process of restricting limitless mental possibility and substituting a quicker but more limited range of expectation and response begins at the back of the cortex and moves forward, not reaching the frontal lobes that control abstract judgment until the brain's owner is over twenty—which justifies why, although your hulking, bristle-faced son has several entirely plausible arguments for lending him and his friends the car this weekend, you are probably wise to say no.[60]

Our children are born, then, as little primates—expecting to spend their next ten years climbing over and around the members of a small, familiar community, learning the right ways to behave from the variety of social situations they must face (all while still small enough not to be able to impose their own tyrannical wills). So if this is the "natural" childhood, what can we do to achieve it? It's not reasonable to assume that you can simply come home from the hospital and declare everyone in your apartment building an allomother—you probably don't know them all yourself. Few jobs welcome children in the workplace, both because children are far too cute for any work to get done and because they are tuned to pay attention to a different, more primitive kind of work: a three-year-old will sit rapt and silent while you make an arrow, but is less interested in how you reconcile last month's receivables.

The fact is that ever since we shifted from hunting to farming—with its long-term worries, heritable ownership, and fixed dwellings—the nuclear family, not the band, has been the theater of upbringing. All we can do is recognize that this represents as much a compromise with our physical natures as is wearing a tie or commuting to work: given the choice, our children would rather live among the Mbuti.

This, though, is to forget that we also have culture—and the wonderful power of culture is how it can adapt to overcome the poor fit between life's current necessities and our hardwired preferences. We cannot (and wouldn't want to) take our children back to the jungle, but we are free to adopt any desirable aspects of primitive culture that might improve our modern life: blurring, for example, the distinctions between child and adult worlds and bringing our kids as much as possible into what interests and pleases *us*, rather than assuming they will be happier in their ghetto of purple dinosaurs. Since learning about life is what childhood is for, isn't it better that our children learn it from us—as they naturally want to—rather than from strangers who are trying to sell them something? Mom and Dad are, in nature, the first celebrities: the famous men we ought to praise are the fathers that begat us.

Which is somewhat surprising, given how society now tends to regard old people: as medical freeloaders, dogs in the manger of inheritance, or pitiful outcasts from productive work. You will remember that people in hunter-gatherer societies tend to reach their peak skill sometime *after* they have passed their peak physical strength; the old are held in honor for what they know and what they remember. They are repositories of history: only they can tell the group whether this emergency or quandary is exceptional or repeated—and if repeated, how it was resolved before. This honor apparently predates modern humans: the Shanidar I skeleton, a Neanderthal man excavated from a cave in the Zagros Mountains of Iraqi Kurdistan, was nearly fifty years old (the equivalent of eighty today), with a broken, withered right arm, a missing right hand, a damaged skull, a severe limp, and heavily worn teeth. It is very unlikely that he would have been able to provide enough food for himself, yet his healed fractures show he was kept alive by his group for years after his injuries.[61] A man of fifty looking for a new job today could only envy him.

Children and old people fit well together, on the principle of "if youth but knew, if age but could." Humans, uniquely among the primates, go through menopause, and one theory is that this is a boost for allomothering: freed from the treadmill of her own breeding, a skilled female can distribute her expertise among the offspring of her younger and less knowledgeable relatives. Old people can be refreshingly rude, both because they know from experience how little some things really matter and because they begin to lose brain function first in the very areas of self-regulation where adolescents acquire them last (so don't lend Grampa the car, either).[62] What they know, moreover, are things you won't get from Wikipedia: stories and conclusions that may not be factually true, but that have been refined and polished through repeated recall into the gems of shared culture. Your grandmother's store of proverbs, parables, genealogies, rhymes—these are the forms of experience to which mere events are just approximations, the promises to a rising generation that, for all the troubles of the day and anxieties of their busy parents, some things yet abide.

NEGOTIATING WITH NATURE

One character looms largest in every hunter-gatherer family, more worthy of attention than even the most respected elder: nature itself. People who depend for survival on food simply showing up at the right moment tend to have a spiritual life involving intense negotiation with the environment. For the Mbuti, the forest itself is the deity, a womb from which they are born and to which they return at death. But she is not simply an indulgent mother. There are all sorts of taboos to observe if she is not to send misfortune along with her bounty—the result is that some 80 percent of the available edible mammals are avoided during some part of their life cycle: the "closed season" is not just an invention of modern game wardens.[63] Almost all hunter-gatherer societies have the practice of leaving behind a little of found treasures, from berries to buffalo. This is effectively insurance: paying a premium into an environmental policy to tide you through the bad times. The Lakota Sioux called the Black Hills of South Dakota "our food store"; Custer's first expedition found gold there—and within two years, twenty-four thousand miners had gone through that pantry like a swarm of ants. It is the old and repeated story: traditional "natives" have learned they must live in balance with the surroundings that support them. Modern profit-takers seek and strip out the carefully tended resources, degrading environment and people alike. This happens so often that there must be a fundamental reason for it.

To be strictly accurate, the "balance" many hunter-gatherer societies finally achieve is not necessarily the one they found at arrival. North America, for instance, was the home of many large and slow-moving animals, from mammoths to giant ground sloths, that did not long survive the appearance of early man. Australian aboriginal people similarly wiped out the larger marsupials and flightless birds. Both cultures routinely burned the wilderness to keep it clear for hunting, much like a medieval king tending his deer park. It has been argued that the true reason humans turned to domesticating plants and animals (beginning, possibly, with figs near Jericho some twelve thousand years ago) is that we had seriously run down the supplies of edible wild things.[64] Happy hunters became unwilling farmers—and there's no question which culture we still prefer: you naturally want to invite your friends around for a barbecue, but not to share a big pot of oatmeal.

The distinction between our exploitative approach to resources and that of our comparatively light-treading ancestors is summed up in the *tragedy of the commons*—a term popularized in a 1968 paper by Garrett Hardin that examined the unequal relationship between private and general benefit.[65] Its classic model is cattle herding in Africa: cows are the measure of wealth for men in many tribes, essential for status and for paying a bride-price. Land, though, is held in common. It therefore makes sense to have as many cows as possible—even thin, sickly cows—despite knowing that overgrazing and desertification are the inevitable result. Personal relative advantage outweighs a general decline.

We should feel no superiority to the cattle herders, because we act as they do in so many cases. Our taste for sashimi, for instance—a style of serving fish that only really caught on, even in Japan, after

World War II—has reduced bluefin tuna stocks by nearly 80 percent in thirty years; they, like several other key commercial fish species, are likely to be extinct in the wild in a few decades.[66] Governments are notoriously bad at regulating fishing because oceans are held in common: others will get the good things if any one country practices restraint. The same applies to minerals, bushmeat, water resources, rain forests, emissions levels—anything left in nature has a lower intrinsic value than what I can extract from it. I have no duty to the all-giving mother, because if it were not me getting mine, it would be someone else.

A tragedy is defined, classically, as the working of inevitability, where character determines fate. We inevitably overexploit our world because our character, our set of instinctive assumptions, is far too local for our current circumstances. A group of 150 people in a large territory, having perhaps monthly contact with a neighboring band, can soon figure out exactly how to live to maintain things as they are, where effort and reward even out. An increase in neighboring population or a reduction in territory introduces competition for resources: simmering low-level war is the usual result, but things overall remain fairly constant—slow-changing, if no longer static. But to have *all* the world's rewards before us for the taking, and all the world's people to supply—that is a situation beyond our capacity to imagine. The solutions we propose to these global problems—the late, emasculated and compromised treaties, the artificial markets in rights to despoil—will always seem unnatural to us because we have no *natural* way of thinking at this scale. Beyond the village, all remains an abstraction.

With our environment as with our child-rearing, with food as with love, doing what comes easily is not necessarily a good choice. The price we have paid for our intelligence as a species is that our history moves much faster than we can evolve: we blithely build new worlds that our old minds find hard to live in. So, if we are not to back ourselves permanently into a corner, we need to admit that our first impulses are not always our best ones. Our instinctive assumptions may be, not just false, but dangerous to our well-being and to our future. What, then, should we do?

We can continue to do what we have always been doing—to adapt—but this time *consciously*. If we have moved seamlessly in so few generations from nomadic tribes of shellfish-eating cave dwellers to fixed, urban, shoe-wearing drinkers of diet Coke, we clearly have the ability to make the necessary cultural changes *now* to live better in the future—rather than waiting until we are forced to make do with some least-bad choice, not living but merely surviving. All it takes to manage such pre-adaptation is foresight, imagination, persuasiveness and confidence—all very human qualities.

THE USES OF CULTURE

Down the track strides a strongly built man, slathered in hog fat, naked save for some bird-of-paradise feathers, a shell nose-ornament, and a strategically placed gourd. He is talking loudly as he comes in a vigorous, warbling language that sounds, to the untutored ear, like water spurting from a pipe. This is Ongka, a headman of the Kawelka people in the Western Highlands of Papua New Guinea; perhaps the only Stone Age person to have written his memoirs.[67]

When Ongka was a young man, his people had no contact with the wider world; they lived tending their sweet-potato gardens with stone axes and wooden paddle-spades, feeding their pigs and fighting their neighbors in endless vendettas over thefts, insults, or witchcraft. The first white men arrived in what Ongka described as "a thunderclap gone mad," landing their plane in the valley below. Everyone was impressed by the lasting quality of the goods they brought—axes that could cut whole trees, knives that did not need constant re-sharpening; many locals assumed that these things had been sent to them by their dead ancestors and somehow appropriated by the white men. Ongka disagreed: "I thought about the problem for some time . . . I thought that we here have never in our *history* had the same powerful new things as the white people, so there could be no real strength in what the people were saying." The new ideas of the Australian administrators were also worth considering: "We must give up this rubbish habit of making war." There was no question in Ongka's mind about losing identity or status: he saw Kawelka culture adapting to new circumstances—naturally, because culture is the center of life.

When the anthropologist Andrew Strathern arrived in the village in 1964, Ongka immediately took him in, despite others' fears that he might be a returning ancestral spirit—because the idea of writing down how people behave seemed an obvious and praiseworthy project. "I taught [him] everything, all the customs of exchange and bride-wealth, all the things about speechmaking, how some people trick and lie, how some tell the truth, how some use concealed forms of talk, I went over all of this again and again for weeks on end." In time, Ongka adopted Andrew as his honorary eldest son, having arranged the appropriate funeral feast and mourning ceremonies when Strathern's own father died.

In 1974, Strathern made a film, *Ongka's Big Moka*, about Ongka's four-year effort to arrange the largest give-away of pigs and possessions ever seen in the region. The *moka* is a ceremony that establishes the status of the giver and lays obligation on the receiver. Here, too, Ongka easily straddled two worlds: his gifts included, as well as pigs and cowrie shells, a Toyota Land Cruiser—and the main recipient was, as well as big-man of an allied tribe, a member of the newly independent Papua New Guinea parliament. Ongka wanted to make his *moka* the biggest on record, because he also knew it would be the last. "The old ways have had it: let us just do this one thing before it all goes quite crazy." At the film's premiere, Ongka came down to the university at Port Moresby and said wistfully that he wished he could add more to his story; so he stayed and dictated his life, smiling and gesticulating at the microphone as he sat among the artifacts in the anthropology lab. "I am just giving my ideas, and I would like them to be known freely, by as many people as possible."

Ongka's self-confident journey into modernity shows that, while we keep the same body through a lifetime and essentially the same genes through all our history, the real evolution takes place outside, in culture. The distinction of nature from nurture is not a useful one—our *nature* is to be socially engaged, interested, and active in the pursuit of . . . anything. We are not condemned to live as our physical mechanisms dictate; in the history of a people as in the lifetime of an individual, we welcome the chance to reshape our circumstances and our expectations—that is what culture is *for.* Naked or clothed, jungle or city dweller, under all skies in all weathers, we are happiest when questing in company. We lose that happiness, though, when we forget the means and concentrate only on the ends—the empty though desired rewards that, like the MacGuffin in a Hitchcock film, merely serve to set the story going.

CHAPTER VII

Living Right

You'd think the most natural role for human instinct would be to help us be good and live right; yet here we habitually take our widest detours into the abstract—we can't prescribe good behavior without using fundamentally untestable concepts. We do, though, seem to have an inbuilt sense of what civility is: it is helpful, sharing, fair, self-restraining. Primate studies show why these qualities are important: life in a band has to balance the efficiencies of hierarchy with the ties of reciprocity; power is necessary, but power within bounds. Primate emotions help regulate these matters swiftly and continuously, but our complex world too often separates the offended from the offender and from the means of retribution. Hence our willingness to commit obvious local wrongs (as in terrorism) in the name of some global right. Hence also our moral axioms, our national characters, our polarized politics: we accommodate the tension between our simple primate emotions and our bewildering world through the connective tissue of culture.

"Tut, tut, child!" said the Duchess. "Everything's got a moral, if only you can find it."

—LEWIS CARROLL, *Alice in Wonderland,* chapter 9

BE good, sweet maid, and let who will be clever," wrote the Victorian moralist Charles Kingsley—and a neat, tripping little phrase it is, highly suitable for a christening mug or old-time sampler. It is only after you read it twice that the puzzle in the motto starts to show through: does being good, then, require no cleverness? Can humans really boast a natural, artless vein of morality that takes no seeking for? Yet Kingsley himself believed the opposite: "For my part, I should like to make every man, woman, and child whom I meet discontented with themselves, even as I am discontented with myself . . . for to be discontented with the divine discontent, and to be ashamed with the noble shame, is the very germ and first upgrowth of all virtue."[1] So being good, for this man at least, clearly did not require being consistent.

Morality may be the oldest topic in human discourse, but it is also the one about which we talk the most guff. Unwilling to admit that our own self-satisfactions and discontents are largely circumstantial—indeed, often the product of blind chance—we call in the air support of Good and Evil to flatten out life's ambiguities. This recourse to higher principles is always problematic and often dangerous, because our sense of morality contains some intrinsic contradictions: it is universal yet personal, important yet vague, the standard of abstract judgment yet steeped in emotion, the foundation of all law yet often merely the custom of our gang—and usually something to be preached more than practiced. It is a mental phantasm, always present, never fully grasped.

The problem of the good has exercised the clever since the earliest days of abstract reasoning—not just as a fine after-dinner topic but also as a basic worry about our standing in life. Are we born good, like the Confucians? Bad, like the Christians? Or simply born (again—unfortunately), like the Hindus and Buddhists? How can we reconcile our genuine, instant admiration for selfless heroism with our inadmissible yet undeniable urges toward the sneaky, the lustful, and the cruel? Why do we find this central goal in life so difficult to explain, except by examples? And what makes the quality of goodness so mutable, from a "good person" to a "good idea," a "good game," or a "good bowl of chili?" All good questions.

Knowing the good presents both conceptual and practical problems. One basic issue, first isolated by the Scottish Enlightenment philosopher David Hume, is that there is no logical route to get from *is* to *ought*: from talking about how people usually behave to the parallel, imagined world of perfection, where all are clothed in the mantles of righteousness.[2] We make this conceptual leap in every lecture we give our children—"you *ought* to know better"—and they would be right to ask us in return: "how do *you* know?" Description is not the same as prescription; the pointing finger is not the thing pointed at.

The other philosophical difficulty with goodness is that it is a fundamentally indefinable term. G. E. Moore, the author of *Principia Ethica*, likened it to "yellowness": you can say that an object reflects light between such and such wavelengths, but if you don't already think of that as *yellow*, the description is no help; and someone looking at the same object might well be seeing what we call "blue."[3] Yet, as with yellow, we all believe we know virtue when we see it. The saint who shares his cloak with the freezing beggar, the soldier who dives on the grenade to save his comrades, the doctor who tends a sick enemy; "thy need is greater than mine"; "I am just going outside and may be some time."

We don't even have to like or agree with those we consider good: Pilate saw the virtue in Christ, Mountbatten in Gandhi, de Klerk in Mandela; many a soldier salutes a "brave and valiant foe" while still going ahead and killing him.[4] Nor is the impression of goodness necessarily dependent on good actions. Can you, for instance, immediately call to mind any remarkable *deed* of the Dalai Lama's, other than his willingness to study quantum mechanics? Yet all who meet him come away with a powerful impression of moral excellence. When, welcoming him after his escape from the Chinese invasion of Tibet, Nehru asked how he was, he replied: "I am quite nice."[5] Perhaps it was not a slip.

The Chinese, of course, would have disagreed—and what is remarkable in our materialistic age is how many deadly disputes are still at least nominally about moral values. One armed party fights for *freedom*; another for *self-determination*; yet another for *the people*; all fight for *justice*. Two hundred and fifty years ago the issue would have been that Frederick wants Silesia and Maria Theresa doesn't want to give it up, but nowadays the governed are happier when their leaders cloak such cynical realpolitik in the language of moral mission. This has not necessarily been an improvement. Where, at the end of eighteenth-century wars, Europe's princely winners and losers could go back to dancing the minuet and trading bons mots, we have no such respite: if evil triumphs, even locally, we cannot forget nor forgive. There is no ethical equivalent of a balance of power.

This presents a puzzle for any student of human nature: while it may be inevitable, even personally advantageous, for humans to be competitive and aggressive, why should they feel the need do it in the name of being peace-loving and restrained? Why should it matter that we think, "I am behaving justly" as we grab our share—or a little bit more? Even more strange, why should most cultures agree that true virtue requires severe discipline and self-denial, that it is as *un*natural as it is necessary? Aspiring to achieve what most find impossible seems an odd result for natural selection—it as if all elephants secretly wished to skate.

Whatever it is, the grammar of morality shows up very early in human life. Six-month-old infants show a distinct preference for a "helper" block that apparently assists another block up a steep incline, while rejecting a "hinderer" that gets in the way.[6] Ten-month-olds register surprise that the climbing block should appear to go to the hinderer for help: doesn't it *know*? The association of good and bad with helpful and unhelpful is strong and natural in children—an effect exploited by Noël Coward when a friend's toddler asked him what two mating dogs were doing. He said that the one in front had suddenly been struck blind and the kind one behind was pushing him to the clinic. The child found the explanation completely satisfying.

MORAL LESSONS FROM THE ANIMALS

Arrive as a guest in any village or encampment around the world and the same thing will happen: someone will offer you food. Sharing good things is the fundamental virtue and by no means confined to humans. Vampire bats will regurgitate blood to feed flock-mates whose hunting was less successful; wolves and hyenas share the kill. In general, social animals that depend on securing rare, high-value nourishment have the habit of then distributing food among not just the helpless children but the group as a whole. It's a behavior that acknowledges the central role of probability: hunting as a pack increases the overall chance of any one animal being successful, but it remains a matter of a chance—today you, brother; tomorrow me.

Probability is what makes ethics both necessary and complicated. If existence were purely deterministic, we could live by simple, hard-and-fast rules: always serve from the left; never contradict your mother. "Even a potato in a dark cellar has a certain low cunning about him which serves him in excellent stead," wrote Samuel Butler. "He sees the light coming from the cellar window and sends his shoots crawling straight thereto."[7] Ants distribute life's complex decisions across so many separate, expendable individuals that each can operate more or less like a robot, integrating sensory inputs to determine a behavioral output. If one of them gets it wrong, she dies—but there are always other closely related ants. No mammal can afford such single-mindedness. Each individual represents too big a parental investment to risk on mere stimulus-response; mammals have to *think*, balancing a collection of ingrained and acquired assumptions about themselves, their companions, and the world. And in doing so, they can display ethical habits that bring them very close to what philosophers claimed as uniquely human territory.

Look at dogs playing in the park. Individuals from breeds as distinct as miniature schnauzers and Pyrenean mountain dogs, strangers to each other, will romp together in a shared routine that looks very much like fighting—but, crucially, isn't.[8] It begins (as it does for coyotes and wolves) with a "play bow," a signal that says, in essence, "what follows next should be considered as friendly"—not an easy message to encode in symbolic logic, but effortlessly understood by any canid. As they play-fight, two dogs are constantly adjusting their behavior, each to suit the other: the Great Dane does not crush the shih tzu, the Scottie does not nip the easily reached tender parts of the Saint Bernard. They are reining in their instinctive aggression, just as the classical moralists would have us do. And if one playmate crosses the line, the response is more than simple pain or anger—there is a note of outrage in the growl, implying a disappointed expectation: *you don't know how to behave.*

Unjustified anthropomorphism? Probably not, if you consider the question of why dogs, or chimpanzees or people, need to play in the first place. The individual's survival depends on membership of the group; the group depends on cooperation; and cooperation depends on an informed knowledge of the intentions and capabilities of others across a range of complex, unpredictable situations.[9] This is too involved a challenge to be met by instinct alone—the *desire* to cooperate may be innate, but the practicalities of actually doing so require learning. Play is therefore a test of fitness for group life: sorting roles to abilities, flushing out the untrustworthy, and establishing dominance (hence the importance of team games at upper-class schools). Most important, play offers a social reward for individual self-restraint. Being a good sport is the first requirement for living in a group.

Of course, the hearty code of the park or lacrosse field is only a beginning. Full social engagement means knowing the right response to a gamut of practical and emotional situations, especially those where the wants of the individual can threaten the successful functioning of the group. Frans de Waal of the Living Links Center at Emory University has devoted his career to studying the moral life of primates; his results show the humbling political skill and subtlety of these cousins we laugh at in zoos.[10] Among golden monkeys, males calm fighting females; among chimpanzees, females reconcile rival males. Individuals trade food and favors, forming friendships and alliances. Baboons grieve for lost companions; rhesus monkeys comfort the depressed. Wild chimpanzees take special care of blind or brain-damaged members of the group. Bonobos defuse tension and build social cohesion with kisses, hugs, and plenty of sex. These may not be techniques applied by, say, the United Nations, but they are all quick and effective methods to neutralize the fissile power of conflict.

Gazing up into the trees or through the lab's one-way glass, we can see three aspects of primate social behavior that could help untangle our own knotty problem of goodness. The first is that all this moral action is also *emotional.* Apes can be devious and calculating (especially in matters of grapes), but they are not coldly rational—or, more precisely, any coldness is itself an emotional statement. When two adolescent female chimpanzees at Arnhem Zoo in the Netherlands refused to come into the building one evening, thus delaying dinner for the whole troop, everyone gave them a thumping for their selfishness the next day: a calculation of offense and punishment, but also a release of anger and resentment that allowed the incident then to be closed. When the Emory researchers brought their chimpanzees big bundles of leafy branches (so useful for so many purposes!), there was a transition in the group's behavior to more sharing and reciprocity, because there was now lots of a good thing to go around. But there was also plenty of hooting and mutual hugging at the prospect, reaffirming the bonds and hierarchies of the group: a celebration of ourselves in good times, like a human Christmas or Carnival.

The second aspect is that primate ethics are *dynamic.* Moral actions don't arise spontaneously from some beautiful Dostoyevskian monkey soul; they address particular threats to the smooth running of the group—and those threats can come from within. As we have seen, the problem of male sexuality is that it ought, in genetic terms, to select for aggressive selfishness—monopolizing breeding opportunities, wiping out the offspring of rivals, and dominating a harem through sheer bulk and bad temper—yet this is not in the interest of females, who benefit from a choice of mates and the opportunity to raise slow-maturing infants in peace. Different primate species resolve the problem differently, but the elements of behavior are the same: female alliances to counter male dominance; peacemaking to reduce male rivalry; grooming, status displays, and other substitutes for mating to satisfy male amour propre; even hidden estrus to muddle questions of paternity. The result of this complex balancing behavior is that everything is kept in play: life is less simple, because genetic and social goods are confounded—so more individuals think they have a reason to remain committed to the group. The peace brought by moral action, therefore, is never static. As in a soap opera, the plot is always changing although the cast remains the same.

The third point about primate ethics is that they depend on an element of the *abstract*—this is what makes them ethics rather than simply responses to stimuli. The constantly shifting emotional balance of group life makes it impossible to treat a particular episode either as unique or as entirely typical; each event is an entanglement of personalities and qualities: "he took my apple" parses into "*it's* not fair" and "*he* is a cheater." The book does not just close on the matter: it might *become* fair if he made public gestures of submission to my righteous rage; he might lose the name of "cheater" if he offered me something good in exchange—but the idea of unfairness and the name of "cheater" remain as conceptual categories to be applied in future cases.

A monkey's friendships and enmities are not determined merely by kinship; relationships of trust and reciprocity depend on actions and responses—things that require long memories, a keen eye for signals, and an ability to separate agent from act, sinner from sin. This is presumably why social animals, from dogs to dolphins, can develop such strong bonds with humans: they have to be good at intuiting how others feel, and this skill can apply even to others from outside their species. The early Russian primatologist Nadia Ladygina-Kohts lived with a young chimpanzee who once escaped onto the roof of her house and refused to come down, even when tempted with his favorite food. Her solution was to sit on the ground and wail as if in pain, at which he immediately rushed back to put his arms around her, gazing into her face with obvious concern.[11] This sensitivity is not restricted to primates: a dog owner will tell you (and research confirms) how upset they act in response to quarrels or distress in their human families.[12] And lest you think this simply an artifact of the pet-owner relationship, remember cats—who Walk by Themselves, and usually respond to a crisis in the house merely by yowling meaningfully near their empty food bowl. It is we, on the contrary, we social primates, who are attentive to *their* feelings.

DO GOOD? DO WELL? FEEL GOOD?

Just how apish are we, then, in the moral sense? Surely some divine spark of reason separates us from the hooting, grinning, dung-flinging jungle dwellers—why else should we have this big prefrontal cortex, with its executive functions, except to free us from the visceral urges of the troop and let us come to logical, defensible conclusions about what is right and wrong? Unfortunately for such an Enlightenment view, all the evidence points in the other direction: we are not just simian in our baser desires, but in the ways we integrate and modulate those desires—right up to the precepts we carve on our hallowed monuments in letters of gold.

Take emotion: you will know from experience that feelings of shame and guilt are often far out of proportion to the actual harm your action has caused. If you were the one child who forgot to bring in your UNICEF contribution to school or the one guest who overslept and missed a friend's wedding ceremony, you will know that "it doesn't matter," while true in abstract terms, does little to cool the puddle of molten lead in the pit of the stomach. Blurting out *danke* when you meant *merci*, using the fish fork for salad—to be in the wrong *hurts*, and the hurt is related to social context. This may explain why, in general, women are more prone to guilt than men—as female primates, they have more at stake in the harmonious running of the group.[13] Men tend to make more of a distinction between observed and unobserved: when no one's looking, you can drink milk straight from the carton.

True, these are minor issues, more manners than morals. All right, then—try this hypothetical question: would you eat your dog? *No!* Well, why not? Protein is protein—people in the Philippines say dog is very tasty . . . *No, it's an intelligent being.* Pigs are even more intelligent, and people eat bacon. If you were really hungry . . . *No! It's my dog, my companion.* OK—say you're starving, and the dog is put to sleep totally painlessly after a happy life . . . *Dog meat would make me sick.* It's fully cooked. *No! It's disgusting!* That, it turns out, is the real point. In studies by Jonathan Haidt using repeated questioning like this, people quickly ran out of defensible reasons to support their strong moral repugnance for practices ranging from the private use of nonaddictive drugs to consensual sibling incest with birth control.[14] They became, in his phrase, "morally dumbfounded": they knew emotionally what they found unacceptable, because it was disgusting, but they had no functional reasoning to justify their disgust. The sense of wrong was intuitive and overwhelming. Haidt also noticed that we make the same queasy faces when considering certain types of moral failing (hypocrisy, betrayal, cruelty, sycophancy) as we do when thinking about feces or rotting meat: they are violations of purity, moral uncleanness. People don't call it "brown-nosing" for nothing; a shit-eating grin is something we all recognize, even though the image is as unlikely as it is distasteful. We are not that far from the dung flingers, after all.

In contrast, Haidt found a strongly shared but equally irrational human response to examples of altruism or self-sacrifice—a feeling he calls "elevation."[15] He quotes from a letter of Thomas Jefferson, who was recommending that a friend read works of uplifting fiction as a kind of workout for his moral sense: "When we see or read of any atrocious deed, we are disgusted with its deformity and conceive an abhorrence of vice," but when the reader comes across any "act of charity or gratitude," it will "dilate his breast"; he will be "deeply impressed with its beauty and feel a strong desire . . . of doing charitable and grateful acts also." Haidt's research found not only a strong approval in many cultures for factual and fictional accounts of altruistic behavior, but he also heard repeated reference to the distinctive feeling of breast dilation that Jefferson mentioned: one Japanese said, "When I see news of a disaster, I feel pain in my chest, and tears actually come out when I read the newspaper. Then after that, seeing volunteers and finding out that there are helping people out there, the pain goes away, the heart brightens up." Others in India mentioned the tingling and warmth that accompany the contemplation of good deeds. We say we feel *touched, moved, inspired* (that is, "breathed into"); physiologists speculate that this may all be the work of the vagus nerve, that mysterious connector of mental and visceral states, but it is clear that, as with disgust, the *feeling* of moral beauty predates and controls any reasoning about the particular act.

This is anecdotal evidence, but there is more direct observation to back it up. Jorge Moll of the National Institutes of Health took fMRI scans of subjects as they decided whether to make a donation from an initial pot of $128 to a range of genuine charities or to keep the money for themselves.[16] Fascinatingly, the same brain region that makes us feel good when receiving money (the ventromedial striatum, which is associated with learning from positive rewards) lights up even more strongly when we hand our money over to charity—it is indeed better to give than to receive. Moll also saw activity in the subgenual area, which is associated with social attachment and the rewards of friendship: doing good doesn't just make us feel good; it makes us feel like a good person. Cleverly, Moll arranged his experiment to include some socially divisive charities, such as those promoting abortion or euthanasia: sure enough, the lateral orbitofrontal cortex (an important region for aversion, anger, and moral disgust) flared up among those who were willing to pay to oppose such causes—their scans glow orange with righteous indignation. Plato had hoped to banish the churning passions down to the thorax, reserving the head for cooler rationality, but they have stayed right up here, shaping our preference before we have even had the chance to articulate it.

And all Moll's subjects, remember, were acting anonymously and alone—his results recorded the individual's interior opinions of social virtue: the pleasures of doing good by stealth. Once we take altruism out into public view, it gains even more emotional power.

Visit any privately funded hospital and you will be confronted by the signs of conspicuous altruism: stepping from the Sam and Irene Weinstein elevator, you stop at the George A. Pappas water fountain on your way to the Alcott Forbes ward. This is not like going to 3-Com Stadium or U.S. Cellular Field, because the labels invite no repayment; Avery Fisher Hall and Dorothy Chandler Pavilion have nothing to sell. They do, though, want us to buy into something: reputation. Which invites the further question, "what is reputation for?"

In primitive terms, the answer is straightforward: mating opportunities. High status implies high resources, making the top ape or human hunter a good bet for supporting the next generation through the dangers of childhood. Humans have developed ingenious ways of storing and displaying their status (cowrie shells, huge stone discs, gold bracelets, private jets), but most cultures have also retained the key status behavior we inherited from the apes: giving things away. Passing out food is key to many primate relationships and the high-status males are usually the most openhanded. Humans, from the ring-giving chieftains in *Beowulf* to Ongka, with his record-beating *moka*, follow in their tracks. The message conveyed is clear: "there's more where that came from."

Are modern humans still subject to this connection between sex and charity? A recent study by Vladas Griskevicius suggests so. His 159 male and female subjects were put into a romantic mood by being shown pictures of smiling, attractive people of the opposite sex.[17] Then they were told they had each received a hypothetical $5,000; they could decide how much of it to spend on a range of luxury goods and experiences, and they could also commit to volunteering some of their time to a range of charitable activities. You will have guessed the result: the men spent the money like water, while the women held on to it. The women volunteered their time to the last minute, while the men remained in selfish isolation. There are two further twists to the results, though: each sex expended its chosen resource—money or time—exclusively in ways that would be visible to others, especially to potential mates. The men spent their money on cars, clothes, and personal adornment, not on vacations or things for the house; the women volunteered to work at central public charitable institutions like homeless shelters, rather than to go clean up a park on their own. Both groups were considering this as an opportunity for display: "Look at me—I have resources to burn." Significantly, the control group of subjects who had not been primed by romantic images did not care about the visibility of their choices.

The second twist is that men will also use benevolence as a mating display—but only via two methods: spending big money and risking their lives. They show almost no interest in contributing time as a demonstration of their nurturing capacities. Thus, among the very rich, you see alpha billionaires like Ted Turner and Donald Trump willingly put up hefty sums in benevolent contributions—although neither is known as a soft touch otherwise—while daredevil tycoons like Richard Branson chance their necks in "charitable" balloon or speedboat stunts. Ex-president Jimmy Carter, by contrast, has spent a quarter century donating his time, intelligence, and experience to a range of difficult problems around the world—and while he may have been effective in this, it has certainly not made him seem sexy.

Far from the sleek and scented world of big giving, there is another arena in which male altruism seems to gain mating opportunities: around the house. A paper from the Council on Contemporary Families shows that, in the forty years since the first significant shake-up of gender roles in the American family, there has been a slow but significant shift in the distribution of work and life responsibilities between the sexes.[18] Male contribution to household tasks has doubled; men have tripled the amount of time they spend on child care; in general, as women take more paying work, men are increasingly picking up the slack in the uncompensated tasks of nurturing and sustaining. Why? Because they get more sex: wives report they have greater sexual interest in a man who does his share of the chores and child rearing.[19] This may be partly because it is easier to feel romantic in an atmosphere of fairness and it's hard to generate much affection for someone you resent, but—at least from the man's point of view—the *reason* for the phenomenon is moot. All men need to know is that being a good husband, like owning a lot of cowrie shells, is a great way to get laid.

Goodness, then, seems to go on three emotional legs: simply feeling good; believing oneself a good person; and looking good as a high-status potential mate. No matter what your philosophy, these emotional cues will inform your decisions and—depending on how you interpret them—determine your abstract sense of right and wrong.

Can we test this through negatives—that is, can we see the effect of detaching emotion from moral judgment? Indeed we can, thanks to our ability to identify very local forms of brain damage. A study that drew together many of the leaders in this field, including Marc Hauser and Antonio Damasio, took six subjects who had suffered lesions in their ventromedial prefrontal cortices (VMPC), the region that informs decisions with emotion.[20] People with such damage normally show a marked decrease in their social emotions—*compassion* or *shame* are just words to them—but their capacity for abstract reasoning is unaffected. The researchers presented them and a control group with a variety of decisions that had little personal and emotional content and found there was almost no difference between the two groups' choices. But when the questions became more emotionally charged and more difficult to resolve ("Soldiers are attacking your village; you and a few others are hiding in a cellar. Your baby starts to cry. Do you smother your baby to preserve your and the others' lives?"), the divergence was immediate. The VMPC patients quickly chose the most "utilitarian" response, totting up lives and choosing the course that preserved the most. They didn't suffer an emotional wrench at the thought of smothering the baby, pushing the one man into the path of the train to save five, or drowning the injured crewman who is overloading the lifeboat. They made, judged by results, the right choice—because they were not troubled by the social emotions of empathy and guilt.

The rest of us, though, remain entangled in the strands of primate emotional affiliation. We applaud generosity in ourselves and others and create artificial dates and rituals of altruism: harvest festivals, Purim charity, tithing, the Islamic duty of *zakat*. We celebrate these local bonds of giving that tie us to our families and communities; and yet we loathe and seek to avoid the one most significant charitable gift we can make each year—when, in support of the old, the sick, the wounded, of schools and hospitals, we pay our taxes. Few people's breasts swell at *that*.

WHY THE GREAT ARE RARELY THE GOOD

"I am the Emperor and I want dumplings!" This was apparently the only coherent command ever uttered by Ferdinand I, emperor of Austria, king of Hungary and Bohemia (as well as ruler of some thirty-three further territories). Dimwitted and severely epileptic, he reigned in unbridled autocracy over a population of nearly fifty million people ("it's easy to rule—the hard part is signing one's name"). As the revolutionary crowds surged toward his palace in 1848, he asked his ministers, "Is that *allowed*?" and, finding he had no power to forbid it, prudently abdicated.

Ferdinand was at least remembered as "the Benign." History offers plenty of examples of rulers almost as unqualified to govern but far less well-intentioned: a shameful parade of tyrants, despots, dictators, strongmen, bullies, lunatics, and butchers. Nor is traditional monarchy the worst offender: the political creeds that most pretended to represent the mass of people, communism and fascism, were also those that slaughtered the largest number of them. Even the U.S. Constitution, painstakingly designed to thwart potential tyranny through its system of checks and balances, still lets through a large number of grasping or corrupt mediocrities. We boast of the system that gave us Jefferson and Lincoln while discreetly forgetting it also produced Richard Nixon and Warren Gamaliel Harding.

"Government," said Thomas Paine, "is but a necessary evil; in its worst state, an intolerable one." Government, though, is also a focus of our obsession with the good; it is the public theater of our morality. Anyone not intending to spend a lifetime in meditation has at least some political opinions—and these are held, not with the light and easy touch of the disinterested, but with partisan vigor and tenacity. Yet despite all our attention and debate, despite five thousand years of experiment with assorted political structures, we have yet to come up with a system that works as smoothly as that of a troop of apes.

Chimpanzees, bonobos, and gorillas, our closest primate relatives, have hierarchical, power-based societies—no Transcendental communitarianism for *them*—but this hierarchy is tempered by resistance. Among all three species, coalitions of the weak can form to restrain, supplant, or even drive out an over-dominating superior. The trigger for resistance is usually unfairness in social exchange (which, as we've seen, is a big issue among social primates), and the force that drives the oppressed to take such a risky step is an equally familiar one: *anger*. Anger is distinct from bloodlust or planned aggression: it is an instant eruptive response ("Hey!" "Oi!" "You!") to a perceived injustice, appropriate for bringing issues out into the open and resolving them quickly. Anger suppressed, as we all know, burns a hole right through the stomach—or worse, gets transferred to some undeserving victim.

There are sound evolutionary reasons to maintain this dynamic balance between the desire of the individual to dominate and the desire of the dominated to resist. In a dangerous world, static groups are sitting targets. Primate bands have power rivalries for the same reasons sensible democracies have term limits: the old ways and faces do not necessarily remain the best. The harsher the environment, the more hierarchical and tense the local politics: baboons, exposed to many predators, have rather a grim Prussian social organization; while bonobos, whose territory spans a particularly lush and productive stretch of jungle, enjoy a more languid, matriarchal, polyamorous culture, earning them the name of "primate hippies."[21]

What is it, though, that drives the individual toward power? What makes one of us wish to dominate the others, given that this risks unpopularity and resistance? "Mating opportunities" is the orthodox Darwinist response—but this is a reason, not an agent. "Serotonin" is probably a better answer: being top ape feels *good*. It provides all the buzz of an adrenaline response without the clammy palms. "Power is the ultimate aphrodisiac," said Henry Kissinger: the connection between conflict, dominance, and sex is very close.[22] Who dares, wins; and who wins, mates (which is why rape is war's inevitable camp-follower).

Your serotonin levels will rise, no matter whether you seize power or have power conferred on you by recognition from the strong or deference from the weak; and, whatever the source, it does things to the brain that alter behavior significantly. Have you, for instance, ever wondered why so many famous and powerful people are such *jerks*? Peter the Great of Russia, on a visit to England, trashed his host's house to such a degree it had to be torn down and rebuilt. Lyndon Johnson delighted in receiving advisers while sitting on the toilet. Prominent people keep everyone waiting, dominate conversation, and make their assistants cry. This is so common that people will say in surprise of a celebrity: "You know, he was actually really nice—like a normal person."

Is this natural selection? Are un-nice guys bound to finish first? Deborah Gruenfeld at Stanford suggests not. Starting with the knowledge that serotonin, as well as making one feel good, reduces inhibitions, she did some controlled studies of the social effects of giving people a little power. For instance, she selected groups of three students at random and set them an artificial task: working on a policy paper. To help fuel them through the collaboration, she put out a plate of cookies.[23] Then one student in each group was randomly chosen to evaluate the work of the other two for a possible cash bonus. Gruenfeld was not interested in the evaluation, though, but in the cookies: the students who had been given the supervisory role not only took more than their fair share of cookies, but they were also far more likely to chew with their mouths open and scatter crumbs on the table. What *jerks*! Other behavioral studies confirm that a little power is bad for the manners. Powerful people—even when their power is only temporary and artificial—are more prone to lounge and slouch, interrupt, contradict, and pick their noses. Less powerful people sit demurely, listen, agree, and cover their mouths when they cough. Powerful people ignore general rules, even those they enforce on others. We've seen balanced-budget congressmen take massive bribes, zero-tolerance governors patronize prostitutes, family-friendly senators seek anonymous liaisons in airport men's rooms—and taxes, we should remember, "are for little people." Such extreme selfishness is not something these people were born with, or they would never have been able to take their first, collaborative steps on the road to power: it is a learned social norm.

Power offers an exhilarating cocktail of reward and risk—and the rest of us seem willing to condone the perks of the powerful as long as they keep those two ingredients in balance, recognizing that they owe us something for the slack we cut them. Henri IV of France paid no attention to hygiene: "His feet and armpits enjoyed an international reputation."[24] He was adulterous, lazy, and self-willed. But he had the grace to acknowledge what the people required of him: that he should rule "with my arm holding the sword and my rump in the saddle," and that he was fighting to assure that every laborer should have "a chicken in his pot every Sunday, at least." We acknowledge that some will be more fortunate, but we demand social engagement from them in return. Thus few begrudged the Roosevelt or Kennedy families their vast inherited wealth, because they willingly assumed the public role that went with it—taking up, so to speak, the sword; planting the rump in the saddle. What we resent is wealth that's kept all for itself and rudeness that dismisses the victim: that's *unfair* and it makes us *angry*.

Our big problem, though, is that unlike our primate relatives, there's not much we can do about it. Anger works quickly in the ape world: offenders are punished, the shrieking dies down, life continues—either as before or with a new balance of power. Our evolved society, however, makes settling conflict a much more difficult business than just giving someone a swift nip. Layers of insulation enshroud the powerful: distance, authority, law, and the cost of litigation separate and protect them from their victims. The favorites they have unfairly elevated have every interest in preserving that unfairness; and society's myriad petty tyrants, itching to exercise any power no matter how minor or obstructive, will be grateful for a chance to browbeat others in the service of "order." It's not surprising the Vienna crowds needed the example of a successful revolution in France before they could work themselves up to depose even feeble Ferdinand.

For many of us, the setting where this tension between primate emotion and human impotence appears most acutely is in the workplace. An office can, unfortunately, combine all the worst aspects of an unhappy family and an oppressive state: favoritism, bullying, and belittlement mix in a toxic stew with pettifogging legalism, corrupt practices, manic competition, and meaningless initiatives. Deborah Gruenfeld's equation of small increments of power with major deterioration in manners plays out in almost every workplace; rare is the employee who has not at some time had to suffer a major jerk for a boss.[25]

This phenomenon seems worst in places where serotonin has the widest play—where power and risk combine most obviously: financial trading floors and hospitals. The Big Swinging Dicks of Wall Street live up to their name because they know they must deliver or die; their lives are a binary function of power and failure. Surgeons who have held a beating human heart in their hands can easily feel how close they are to gods. Meanwhile, management will find it hard to rein in these star performers just because they have no manners: the occasional thrown telephone or groped bottom seems a small price for securing such rare skills. Besides, most managers themselves rose through the same system: they've been chewing with their mouths open for years.

The result is some very bad places to work, where emotional poison gets passed around in penny packets from those who can dish it out to those who have to take it. In 2003, the journal *Orthopaedic Nursing* reported that 91 percent of nurses had suffered verbal abuse within the previous month; in a recent study of three thousand medical students, 42 percent mentioned being harassed, and 84 percent said they had been belittled.[26] Just a bracing climate of healthy criticism? No, not really: an alarming paper by Amy Edmondson of the Harvard Business School reveals that nurses in units with a demeaning or hypercritical culture are up to ten times *less* likely to report any errors in drug treatment.[27] Bad manners can be fatal—but not, unfortunately, to the bad-mannered.

Henri IV's sincere curiosity about his people and closeness to their concerns served a greater purpose than simply making him seem responsive and fair: it also prevented him from being blinded by the ceremonies of power. The insulation our society gives to the prominent not only protects them from our indignation when they behave like jerks, but it can even keep them from knowing they are jerks in the first place: who would be willing to tell them? Surrounded by the rewards of success, they lose their fear of risk and their interest in conflicting opinions. There's a revealing story in Bob Woodward's book *State of Denial*: General Jay Garner was being debriefed by the Bush administration after his term as head of "postwar" planning in Iraq in June 2003. He felt that three disastrous mistakes had been made: disbanding the Iraqi military, outlawing the Baath Party, and replacing an Iraqi-fronted government with an American proconsul—but that all three were reversible, if addressed promptly. When Garner found himself in the Oval Office, though, he could see that this was not a message anyone cared to hear; he found himself talking, not about the crisis-in-making, but about how personally beloved the president was in Iraq. Bush, in response, joked that perhaps Garner's next assignment would be the invasion of Iran. Faced with the incurious egotism of power, Garner lost the nerve to speak truth to it: "I think if I had said that to the President in front of Cheney and Condoleezza Rice and Rumsfeld in there, the President would have looked at them and they would have rolled their eyes back and he would have thought, 'Boy, I wonder why we didn't get rid of this guy sooner?'" Garner sums up: "They didn't see it coming. As the troops said, 'they drank the Kool-Aid.'"[28] The general had not drunk the Kool-Aid—but, in moral terms, he had eaten his own dog.

You may never yourself be in the position to set the president straight, but you will be bound to encounter plenty of jerks elsewhere whose sense of fair play has been damaged by power. The best advice in dealing with them seems to be to remember that their offenses and your anger are both residues of a simpler, face-to-face life in the forest. At the most basic level of human civilization, among the hunter-gatherers, an awful lot of effort goes into leveling social relations, by insisting on sharing resources and enforcing taboos against boastful or bossy behavior. These "primitive" people are acutely aware of the potential dangers of hierarchy and consciously resist it. You should act, therefore, as the inner tribesman would have you do: point out immediately, clearly, and calmly how such and such action is unfair, belittling, or rude. Insist that this is *wrong* (for if it isn't, what is?) and that you will not accept it. Warn that if it is repeated, you will certainly seek to have it publicized and punished. In all this, you show your power—over words, your own emotions, and the situation—sending out a signal that can sober up even the most serotonin-intoxicated. We live in an unfair world and this may not always work, but at least you will have done (as any chimpanzee would agree) the right thing.

KILLING FOR PEACE

"Our words are dead until we give them life with our blood." The voice is soft, a thick Yorkshire burr giving it the tone of a well-played but homebuilt instrument. "Your democratically elected governments continuously perpetuate atrocities against my people all over the world. And your support of them makes you directly responsible, just as I am directly responsible for protecting and avenging my Muslim brothers and sisters. Until we feel security, you will be our targets."[29] The speaker on the videotape is Mohammad Sidique Khan. Born, brought up, and educated in the northern English city of Leeds, he worked in a local elementary school as a "learning mentor" for immigrant children. He was known locally as a kind and caring man with "100 percent commitment," whose self-appointed role in the community was somewhat like a camp counselor's: organizer of sports, go-between in disputes with authority, guide and friend to troubled youths. He was also the leader of the suicide gang that exploded four bombs on London Transport on July 7, 2005, killing fifty-two people and injuring seven hundred.

What injustice had Khan personally suffered? Essentially, none: he had close relations with many non-Muslim colleagues and friends, whom he asked to call him "Sid." He received grants from government agencies and charities for setting up gyms; his mother-in-law, also prominent in local education, had been invited to Buckingham Palace and met the Queen. No, what motivated Khan was the *image* of suffering: the "the bombing, gassing, imprisonment and torture of my people" by governments with "power and wealth-obsessed agendas." He was a man who liked to empathize, to make things happen, and to earn praise and respect. He was also the father of a small child—with another on the way—and was not, perhaps, making quite as big a mark on the world as he had hoped.

Diffuse, general anger at arrogant injustice; a gnawing sense of personal powerlessness; intense but free-floating feelings of family loyalty and sympathy with the suffering—these are what make someone a terrorist footsoldier. They are also the essential elements of indoctrination for suicide bombers. Marc Sageman of the Foreign Policy Institute found that the great majority of the nine hundred al-Qaeda operatives he had studied were, like Khan, part of a warm family, well educated, usually married with children.[30] They tended to have studied science, not religion—which suggests they were people who had once hoped to gain prestige by changing the world, doing something important for their people. Sageman also says that the ties that draw new recruits into terrorist cells tend to be personal, not ideological. The majority are already friends of current members, people they had eaten with, played soccer with, whose mothers they knew; Khan's group fondly addressed each other as "bruv." Personal loyalty—not letting down one's side—thus adds a powerful spur to going through with a suicide mission. In close-knit communities, moreover, the "noble" dead become celebrities, their faces appearing on walls like sports stars or entertainers; they have achieved in one blinding moment the fame that would have escaped them in a lifetime of obscure toil. The motivation of the terrorist thus seems less bloodthirsty nihilism than the sentimental resignation and self-pity of Sydney Carton in *A Tale of Two Cities*: "It is a far, far better thing I do than I have ever done. It is a far, far better rest I go to than I have ever known."

Our emotions, bred in the distant past of intimate tribal life, offer no clear guide for keeping apart personal from general frustrations, local from global sympathies. Thus a Saudi, paid by a Syrian, can kill fifty Iraqis in the sincere desire to punish an Israeli for oppressing a Palestinian. He has no reasoned idea of the political purpose of his actions—and still less of the humanity of his victims; he is striking in righteous wrath at an imagined tyrant to gain the gratitude of a fictional family. It may be pointless, cruel, wasteful—it is not, sadly, inhuman.

What *is* inhuman is the effect of prolonged and complete powerlessness. You will remember from chapter 2 how universal was the human wish to participate altruistically in a group, enforcing fairness by punishing cheaters. Well, it's only *nearly* universal: hunter-gatherers have it; North American students have it; elderly ex-Soviet Russians, however, and indigent peasants from Uttar Pradesh, India's poorest province, do not.[31] They instead will willingly give up money to punish *altruists*, or just to keep down someone else. Why? Perhaps the empty platitudes of socialist brotherhood soured the Russians against "do-gooders" of all kinds; the Indians simply ascribed it to jealousy or said, "Imposing a loss on him gives me enjoyment." This implies that if the levers of social adjustment lie dead in your hands long enough, you can forget what they are for and start whacking people at random because, at least, it's better than being whacked. Of all the arguments for the extension of democracy, this seems the most natural: unless people have the chance to use power to regulate their circumstances, they will come to see power purely as the medium of pain and will seize any opportunity, no matter how senseless, to inflict it. Adlai Stevenson, here as on so many topics, put it best: "Power corrupts, but lack of power corrupts absolutely."

STRIVING FOR THE ABSOLUTE

Like the physical universe, moral space is curved: you can travel as far as you like, but you will eventually return to David Hume. We inevitably run up against his paradox because, as social animals, we have no choice but to think in terms of *ought*, when we only have experience of *is*. If we are to act based on that experience, we *have* to pretend it is general, recasting the vivid moment into abstract laws, taboos, commandments, and homely proverbs—none of which we can be certain of, for they are not absolutely true. They serve merely, at best, as a guide to what usually happens; and what usually happens will usually change.

Try, for instance, explaining the seven deadly sins to a caveman. You will find it hard going—not because he has no moral compass, but because his compass is set skew to the coordinates of medieval European life, whence these sins derived. Lust, for example, is not a great source of trouble for most hunter-gatherers: sexual thoughts are instead the engine that moves the tribe's story forward. Only in a crowded landscape does wayward desire break homes and hearts. Gluttony? You must be joking: we eat all we can. Only farmers need deny themselves through the hungry spring to assure enough seed for planting. Covetousness does not matter for people with few possessions. Sloth? Again, a concern only for those with rows to hoe and a long season to toil before the rewards of harvest. Pride, though—*that* makes sense: no social unit can long survive if some people think themselves too good for the others; if you were a missionary to the past, you could start the conversation there. Yet the present, too, can pose problems for the old laws: the Vatican itself has recognized that sins adjust to time and place. In 2008 it mooted a list of new, "globalized" mortal sins as an appendix to the familiar seven: it includes genetic experimentation, polluting the environment, taking drugs, and being obscenely wealthy.[32] These, in turn, might take some explaining to a medieval peasant.

Sometimes, though, our abstractions can provide a useful tool for altering moral garments to fit a changing body of behavior. In Jewish law, there is the concept of *geneivat da'at*, or "theft of mind," derived originally from a grammatical dispute about the eighth commandment (or seventh, for Catholics and Lutherans): thou shalt not steal.[33] Over the millennia, this has come to be interpreted as a general prohibition against *creating a false impression for gain* or *acquiring undeserved goodwill*. From Old Testament matters like offering expensive anointing oil from an empty bottle (knowing it will be politely refused), it has mutated to cover such up-to-the-minute sins as deceptive advertising, cheating on exams, improper auditing practices, and mislabeling clearance sales as discounts—all sins that the devout must shun as they would the original theft in the Decalogue. The compilers of the Babylonian Talmud might be surprised at these applications, but they would recognize immediately the usefulness of stretching a precept while trying not to break it. Legal tradition, at least in the English-speaking countries, shows a similarly inventive flexibility: most modern contract law is simply an ingenious extension of the rules about hiring someone's horse.

Any law or commandment, though, even the most revered, is circumstantial. What we would prefer are *principles*, moral concepts that link the way we actually behave with the purer terms of philosophy. In the 1970s, Lawrence Kohlberg at Harvard set out in search of them;

a pragmatic optimist, he followed the upward footsteps of the Swiss developmental psychologist Jean Piaget, tracking the moral progress of the individual from the willful atom of infancy to the self-regulating citizen of maturity. In Kohlberg's view, the key to the process was *abstraction*, achieved through ascending levels of moral development. Over time and confirming experience, our minds learn to separate good actions from reward, bad from punishment; then to transcend mere social exchange; then to leave behind those impostors, praise and blame, law and disorder, even to discard the fiction of the social contract—and to know at last Good and Bad so purely that they deserve their capital initials, standing face to face with them at the moral summit where we do things for no other reason than that they are Right.[34]

Kohlberg tested his subjects' ethical levels using dilemmas like this one: your spouse is dying of a rare disease.[35] The druggist has discovered a cure, but it costs more money than you have and he is unwilling to lower the price. Do you break into his store at night to steal the drug? Kohlberg didn't care whether the answer was yes or no, but what the subject's justification was. Level 1 would say, "No, I might get caught," or "Yes, because I really need it." Level 6, however, would understand implicitly that this was an abstract question, weighing the worth of human life against the right to property and would judge them thus—as absolutes.

Kohlberg hoped to find that at least some Harvard men had reached this final stage—and yes, he tested mostly men. When another Harvard psychologist, Carol Gilligan, interviewed a range of women, however, she found most of them resolutely refusing to stick to Kohlberg's terms and insisting that each situation had its own dynamics: "How well do we *know* the druggist?" was one typical response.[36] In Gilligan's view, this showed there was no unique value in progressive abstraction toward an impersonal standard of justice. Emotional engagement and empathy—moral qualities that she called the "ethics of care" and ascribed preferentially to women—can make us just as good without resorting to capital letters.

If different genders can have their characteristic ethics, then, what about different nationalities? Tradition assumes that morality differs by country, and various untranslatable concepts suggest that tradition is correct: Swiss *leumund*, the court of local reputation, is clearly different from French *gloire*, the luminous primacy of the state. The Serbs have *inat*, a version of spite so thorough that it extends from "rather than surrender, I will eat grass" to "my greatest hope is that my neighbor's cow will die." The Danes suffer under *Janteloven*, a grim communal disapproval of anyone thinking himself special in any way (even by giving his child an unusual name, which is actually illegal). *Pashtunwali*, the code of the Pashtun people of Afghanistan, severely enforces rigid principles of honor, revenge, bravery, pride, and protection—all summed up in *nang*, a mystical idea of purity of reputation. "I despise the man who does not guide his life by *nang*" cried the seventeenth-century poet Khushal Khan Khattak. "The very word *nang* drives me mad!" The West, now camped in Pashtun lands, cannot afford to ignore *nang* either.

The national stereotypes we devise about each other recognize unique virtues as well as vices in other peoples. In heaven, we hope, the cooks will be French, the policemen English, engineers German, lovers Italian, and bankers Swiss (it is in hell that the cooks are English, policemen German, engineers French, lovers Swiss, and bankers Italian). We like to think that these national virtues automatically translate into personal excellences—that an Irish passport guarantees eloquence or Dutch one tolerance—but the facts are different. A study in 2005, covering forty-nine separate cultures and conducted by the National Institute on Aging, found that while stable national stereotypes are widely held, both about other groups and one's own, these are entirely inaccurate when compared with the mean results of personality tests.[37] That is, we are not only wrong about others; we are wrong about ourselves. Canadians are *not* less assertive, more altruistic, and more modest than citizens of the United States—they only think they are. French Swiss, who claim to be among the most introverted of peoples, are more extroverted than the Spanish, who give themselves a high rating for this quality. Easygoing Chileans are just as conscientious as uptight Germans. And Czechs, who consider themselves very disagreeable, are in fact nicer on average than all the top five self-rated countries. And what nation, do you think, rates among the highest for adventurousness, extroversion, and general liveliness? Yes, the reserved, aloof, stiff-upper-lip English. It seems national characters are like national dress—something only put on for the tourists.

The fiction of shared character becomes obvious when you talk with, or about, individuals—because there is one moral ideal we all have that does not derive from proverbs or commandments, from culture or education, nor even necessarily from our primate heritage: our own free will. We would rather be good through choice, not from compulsion. We value self-control in part because it is done by the self, not imposed by authority or social pressure. Obedience and conformity are marks of weakness; discipline and restraint reveal strength. Thus we find, whenever we consider the good, a confusion between what's good for the group (our values, our traditions) and our personal conviction that it is best to be free—and this confusion deepens when we remember that, if freedom is best for one, it must be best for all, even those whose traditions we despise.

The Declaration of Independence proclaimed our inalienable right to liberty as a *self-evident* truth—an axiom that needs no proof. Yet Thomas Jefferson, the declaration's author, would have been the first to admit that this was a fudge—a political compromise agreed to paper over some deep disagreements about what such "rights" really meant. The historian David Hackett Fischer has pointed out that the term "liberty" had crucially divergent meanings in the four quarters of the new American Republic.[38] Liberty was the freedom to enact the will of God in puritan New England; to deal fairly in Quaker Pennsylvania; to live according to your station in the gentlemanly Old Dominion; or to do whatever you damn well please in the wild Appalachians. One word, but four separate ideals. Even the founding fathers, heirs to the clear-eyed Enlightenment, found the virtuous life a difficult thing to define. Nowadays, we see the same confusions, the same blurring of definitions when we try to reconcile the high-sounding abstractions of political life with our passionate local loyalties: we commit ourselves to general slogans or reduce our tests to single issues. Institutional morality, even the most self-consciously rational, fits real life poorly: the fact is, we will always have a more sophisticated and balanced appreciation of our sports teams than of our governments.

TO END AT THE BEGINNING

A world in which people were always right would clearly be a fiction—and a pretty dull one at that. *Being* right, as we've seen, is itself largely fiction: a brave attempt to impose consistency and purpose on a life that remains stubbornly random. Even the simplest sense we make of what our senses tell us is a fiction, composed by a brain designed, above all else, to simplify experience by imbuing it with apparent meaning. What, then? Are we doomed to have all our accomplishments marked with an asterisk? Must we live on forever as *Bozo sapiens*, the wise fool?

To approach an answer, it is worth recalling what being right is *for*. What is the common sense that, in the many examples that dot these pages, we have lost when we head off into error? In essence, it is a sense of *probability*, of the complex relations between events and rules, patterns and causes. In our lifelong journeys, we humans tend to navigate like coasting sailors, not transoceanic pilots: we look out for landmarks and invent rules of thumb. Our minds, so acute locally but woolly in general, try to concoct the best possible sense out of what life shows us here and now, rather than to develop a consistent picture of how everything fits together. Thus we are built to be interested in and judicious about the incidents and quirks of what we know well—*that* line of surf over the reef, *that* darkening of the upwind horizon—while our broader explanations so easily shade off into krakens and mermaids.

In the simpler past, this was probably a sensible arrangement, because without an explanation—any old explanation—the world becomes worrisome; and existential worry is a needless distraction from more pressing matters of survival. Only now has our existence become too complex for such an easy cohabitation between the thoroughly known and the merely guessed at. Life doesn't separate neatly: local and global are intermingled, and each is compounded of elements both deterministic (like technology, taxes, and the law) and essentially random (like politics, finance, sickness, and love). We are asked to respond to the world at every scale, and have to be as interested and informed about the widest matters as we are about our most personal expertise. We are no longer just explorers, coming to provisional conclusions as we go; we are custodians. We are responsible.

In this, we are like Dietrich Dörner's kings of Tanaland, so it is worth remembering what set apart the few successful ones from the many failures: they thought *probabilistically*, admitting the power of the random and the unknown. They took steps one by one, testing the responses of a complex system they did not pretend they had mastered. They retained the primordial human urge to examine new things closely. They did not, for no human can, comprehend the situation in all its fearsome nonlinear intricacy—but they used the straight lines of local conclusions to approximate the wider curves of probability.

The other hope for damping down our all-too-human capacity for error is to recognize and harness our equally human talent for making fictions, embodying an indifferent world in forms that let us engage with it: pictures, models, metaphors, philosophies, aspirations. We will always have many ideas; not all of them need be stupid. It's not a *fault* that scientists don't think "scientifically" when they seek their inspiration; it's an essential part of discovery. They could not truly care about their field without some *un*scientific mental image of the beautiful, mysterious thing they seek through their rigorous and self-disciplined method. The human urge to create fictions—that is, culture—gives us the tools to shape events, while at the same time subtly reshaping our own expectations. It motivates us to go out and create new explanations, because we find such creation a pleasure. These fictions make our fortune—for if we lacked the ability to rewrite our story as we go along, we would doubtless have died out long ago . . . victims of our own certainty.

During the initiation ceremonies of the Zuñi, a Pueblo people of the Southwest, the moment comes when the boys, having been whipped one last time by the "scare kachinas," masked punitive gods they have learned to fear from infancy, must stand still and face them—to then see the masks lifted and the features of their fathers and uncles revealed.[39] Once sworn to secrecy, they will put on the masks themselves, take up the yucca whips, and beat their elders in turn. There could be no more powerful way to convey what it means to reach maturity.

Many of the grand abstractions we attempt to apply to life—free will, allegiance, justice, truth—are just such scare kachinas: neither entirely god nor man, neither divine power nor personal whim. They are responsibilities we must take on, in the full knowledge that they will always be greater than ourselves. It is the same with all the species of error we've looked at through this book: they are the permanent companions of our capabilities. To err is human because we could not *be* human without error. We err because we seek, we fail to grasp because we try the furthest reach. Our tale is told neither through theology nor biology, but through history—a discipline in which we are not allowed to lie, yet can never prove our conclusions true. We take a world of randomness and try to mold it into meaning, using a mind suited best to the immediate and the familiar. We talk of powers and dominions—but we think in the here and now, this chilly night, watching the loved faces emerge from behind the masks.

Notes

CHAPTER I: FROM THE LOGBOOK OF THE SHIP OF FOOLS

1. FUBB: " . . . beyond belief"; FUBAR: " . . . beyond all recognition."
2. Carlo M. Cipolla, "The Basic Laws of Human Stupidity," *Whole Earth Review*, Spring 1987, pp. 2–7.
3. Aristotle, *On Sophistical Refutations*, trans. W. A. Pickard-Cambridge, bk. 1, ch. 1, in *Works of Aristotle*, Oxford University Press, Oxford, 1928.
4. See the works of Ann Coulter.
5. Reported from a telephone survey; see http://thenonsequitur.com/?p=335.
6. Gilles Deleuze, quoted in Richard Dawkins, "Post-Modernism Disrobed," *Nature*, July 9, 1998.
7. Personal experience—unfortunately.
8. Harry G. Frankfurt, *On Bullshit*, Princeton University Press, Princeton, NJ, 2005.
9. Francis Bacon, *The Works of Francis Bacon*, vol. 3, ed. J. Spedding, R. L. Ellis, and D. D. Heath, Longmans, London: 1887, p. 283.
10. Ibid., *Novum Organum*, 1620, bk. 1, sec. 95.
11. Described in Voltaire, "De l'Encyclopédie," *Oeuvres complètes*, vol. 22, Hachette, Paris, 1860, pp. 357–59.
12. Immanuel Kant, "Was Ist Aufklärung?" *Berlinische Monatsschrift*, ed. J.F. Biester, F. Gedike, Berlin, December 1784.
13. Bacon, *Novum Organum*, 1620, bk. 1, sec. 49.
14. Cited in D. B. Owen, ed., *On the History of Statistics and Probability*, Marcel Dekker, New York, 1976.
15. Quoted in Robert Robinson, *Skip All That*, Arrow Books, London, 1997, p. 197.

CHAPTER II: IDOLS OF THE MARKETPLACE

1. R. Bshary and R. Noë, "Biological Markets," in *Genetic and Cultural Evolution of Cooperation*, ed. P. Hammerstein, MIT Press, Cambridge, MA, 2003.
2. R. Bshary and A. Gutter, "Image Scoring and Cooperation in a Cleaner Fish Mutualism," *Nature*, vol. 441, June 22, 2006, pp. 975–78.
3. N. J. Emory and N. S. Clayton, "Effect of Experience and Social Context on Prospective Caching Strategies by Scrub Jays," *Nature*, vol. 414, November 22, 2001, pp. 443–46. They are quoted in "Don't Call Me Birdbrained," *New Scientist*, June 23, 2007, p. 35.

4. S. P. Henzi and L. Barrett, "Infants as a Commodity in a Baboon Market," *Animal Behaviour*, vol. 63, May 2002, pp. 915–21.
5. Michael L. Platt and Paul W. Glimcher, "Neural Correlates of Decision Variables in Parietal Cortex," *Nature*, vol. 400, July 15, 1999, pp. 233–38.
6. M. Keith Chen, Venkat Lakshminarayanan, and Laurie R. Santos, "How Basic Are Behavioral Biases? Evidence from Capuchin Monkey Trading Behavior," *Journal of Political Economy*, vol. 114, no. 3, 2006, pp. 517–37.
7. G. Belsky, "Americans and Their Money," *Money*, vol. 22, no. 3, March 1993, pp. 114–17.
8. Adam Smith, *Wealth of Nations*, Strahan and Cadell, London, 1776, bk 1, ch. 2, para. 2.
9. Calculation correct as of July 2007. Your figures may vary. Past performance is no guarantee of future performance.
10. L. J. Savage, *Foundations of Statistics*, Wiley, New York, 1954, p. 103.
11. William Beckford, *The History of the Caliph Vathek*, J. Johnson, London, 1786, ch. 1.
12. Ibid.
13. The calculation of Beckford's fortune is based on Dominic Webb, "Inflation: The Value of the Pound, 1750–2005," House of Commons Library Research Paper, February 13, 2006.
14. Quoted in Boyden Sparkes and Samuel Taylor Moore, *Hetty Green: The Witch of Wall Street*, Doubleday, Doran, Garden City, NY, 1935, p. 120.
15. Quoted ibid., p. 139.
16. Camillo Padoa-Schioppa and John A. Assad, "Neurons in the Orbitofrontal Cortex Encode Economic Value," *Nature*, vol. 441, May 11, 2006, pp. 223–26.
17. Platt and Glimcher, "Neural Correlates."
18. This paragraph draws largely on Brian Knutson, Scott Rick, Elliot Wimmer, Drazen Prelec, and George Loewenstein, "Neural Predictors of Purchases," *Neuron*, vol. 53, January 4, 2007, pp. 147–56.
19. Most notably Antonio Damasio, now at the University of Southern California.
20. Knutson et al., "Neural Predictors."
21. P. A. Samuelson, "Risk and Uncertainty: A Fallacy of Large Numbers," *Scientia*, April–May 1963, pp. 1–6.
22. A. Tversky and D. Kahneman, "Loss Aversion in Riskless Choice," *Quarterly Journal of Economics*, vol. 106, no. 4, November 1991, pp. 1039–61.
23. Chen et al., "Behavioral Biases."
24. Daniel Kahneman, Jack L. Knetsch, and Richard H. Thaler, "Anomalies: The Endowment Effect, Loss Aversion, and Status Quo Bias," *Journal of Economic Perspectives*, vol. 5, no. 1, Winter 1991, pp. 193–206.
25. Ziv Carmon and Dan Ariely, "Focusing on the Forgone: How Value Can Appear So Different to Buyers and Sellers," *Journal of Consumer Research*, vol. 77, December 2000, pp. 360–70.

26. Amos Tversky and Daniel Kahneman, "The Framing of Decisions and the Psychology of Choice," *Science*, vol. 211, January 1981, pp. 453–58.
27. Terrance Odean, "Are Investors Reluctant to Realize Their Losses?" *Journal of Finance*, vol. 53, no. 5, October 1998, pp. 1775–98.
28. RCA hit a high of $114 per share in September 1929; it never regained even this nominal value, eventually being acquired by GE in 1987 at $66.50 per share.
29. "Stockmarket Extremes and Portfolio Performance," Towneley Capital Management, 1994.
30. This effect was vividly demonstrated in a study by Brad M. Barber and Terrance Odean: "Trading Is hazardous to Your Wealth: The Common Stock Performance of Individual Investors," *Journal of Finance*, vol. 55, no. 2, April 2000, pp. 773–806, which showed an across-the-board *negative* correlation between volume of trading and net return for individual investors at a discount brokerage.
31. W. Samuelson and R. J. Zeckhauser, "Status Quo Bias in Decision Making," *Journal of Risk and Uncertainty*, vol. 1, no. 1, 1988, pp. 7–59.
32. Kelley Blue Book, values for a 2002 Ford Taurus.
33. Personal experience, 1972 Oldsmobile Delta 88.
34. Jennifer S. Lerner, Deborah A. Small, and George Loewenstein, "Heart Strings and Purse Strings: Carryover Effects of Emotions on Economic Decisions," *Psychological Science*, vol. 15, no. 5, 2004, pp. 337–41.
35. A partial description of the Mader house in North Andover, Massachusetts, in Linda Matchan, "XXL: Just How Do You Furnish a McMansion?" *Boston Globe*, February 9, 2006.
36. Richard Austin Smith, *Corporations in Crisis*, Doubleday, Doran, Garden City, NY, 1963, p. 64.
37. The Boeing V-22 Osprey tilt-rotor aircraft cost ten times its original budget. Source: Christopher Bolkcom, "V-22 Osprey Tilt-Rotor Aircraft," Congressional Research Service report for Congress, March 13, 2007. Kansai airport is sinking.
38. Ralph Atkins, "The Car Club That Takes the Fast Lane," *Financial Times*, June 30, 2007.
39. Gregory S. Berns, Jonathan Chappelow, Milos Cekic, Caroline F. Zink, Giuseppe Pagnoni, and Megan E. Martin-Skurski, "Neurobiological Substrates of Dread," *Science*, vol. 312, no. 5774, May 5, 2006, pp. 754–58.
40. Camelia M. Kuhnen and Brian Knutson, "The Neural Basis of Financial Risk Taking," *Neuron*, vol. 47, no. 5, September 1, 2005, pp. 763–70.
41. If you think this characterization unfair, consider the fact that medication and other therapies for Parkinson's can quickly turn sober, cautious people into compulsive risk-seekers. For a thoughtful discussion, see Michael J. Frank, Johan Samanta, Ahmed A. Moustafa, and Scott J. Sherman, "Hold Your Horses: Impulsivity, Deep Brain Stimulation, and Medication in Parkinsonism," *Science*, vol. 318, November 22, 2007, pp. 1309–12.

42. Antoine Bechara, Hanna Damasio, Daniel Tranel, and Antonio R. Damasio, "Deciding Advantageously Before Knowing the Advantageous Strategy," *Science*, vol. 275, no. 5304, February 28, 1997, pp. 1293–95.
43. Sabrina M. Tom, Craig R. Fox, Christopher Trepel, and Russell A. Poldrack, "The Neural Basis of Loss Aversion in Decision-Making Under Risk," *Science*, vol. 315, no. 5811, January 26, 2007, pp. 515–18.
44. The term originates in George Ainslie, "Specious Rewards: A Behavioral Theory of Impulsiveness and Impulse Control," *Psychological Bulletin*, vol. 82, no. 4, 1975, pp. 463–96.
45. This is a simplification of the conclusions from A. Raineri and H. Rachlin, "The Effect of Temporal Constraints on the Value of Money and Other Commodities," *Journal of Behavioral Decision-Making*, vol. 6, 1993, pp. 77–94.
46. Samuel M. McClure, David I. Laibson, George Loewenstein, and Jonathan D. Cohen, "Separate Neural Systems Value Immediate and Delayed Monetary Rewards," *Science*, vol. 306, October 15, 2005, pp. 503–7.
47. He mentioned this, apparently, in an interview with *L'Humanité* in 1971; quoted in "Four on the Road" *Time*, November 8, 1971.
48. The company is called StickK (www.stickk.com); its success will be a good indicator of our self-control.
49. John D. Beaver, Andrew D. Lawrence, Jenneke van Ditzhuijzen, Matt H. Davis, Andrew Woods, and Andrew J. Calde, "Individual Differences in Reward Drive Predict Neural Responses to Images of Food," *Journal of Neuroscience*, vol. 26, no. 19, May 10, 2006, pp. 5160–66.
50. See W. K. Bickel and M. W. Johnson, "Delay Discounting: A Fundamental Behavioral Process of Drug Dependence," in *Time and Decision*, ed. G. Loewenstein, D. Read, and R. F. Baumeister, Russell Sage Foundation, New York, 2003.
51. W. T. Brannon, *The Con Game and "Yellow Kid" Weil*, Dover, New York, 1948, p. 294.
52. Michael Lewis, "Jonathan Lebed: Stock Manipulator, S.E.C. Nemesis—and 15," *New York Times Magazine*, February 25, 2001.
53. Much of the information in this section comes from Nicholas Barberis and Richard Thaler, "A Survey of Behavioral Finance," ch. 18, in *Handbook of the Economics of Finance*, ed. G. M. Constantinides, M. Harris, and R. Stulz Elsevier Science, Amsterdam, 2003; and Sendhil Mullainathan and Richard Thaler, "Behavioral Economics," MIT Department of Economics Working Paper Series, September 27, 2000.
54. NYSE data, July 14, 2007.
55. Gaurav S. Amin and Harry M. Kat, "Hedge Fund Performance 1990–2000: Do the 'Money Machines' Really Add Value?" *Journal of Financial and Quantitative Analysis*, vol. 38, no. 2, June 2003, pp. 251–74.
56. W. De Bondt and R. Thaler, "Does the Stock Market Overreact?" *Journal of Finance*, vol. 40, no. 3, 1985, pp. 793–808; Paul Tetlock, "Giving Content to

Investor Sentiment," *Journal of Finance*, vol. 62, no. 3, June 2007, pp. 1139–68; Tom Arnold, John Earl, and David North, "Are Cover Stories Effective Contrarian Indicators?" *Financial Analysts Journal*, vol 63, no. 2, March–April 2007, pp. 70–75.

57. The artist is Pietro Manzoni, who sold ninety of these cans, each one ounce, for the current price of gold in 1961. The Sotheby's sale price for one can in 1991 was sixty-seven thousand dollars. John Miller, "Excremental Value," *Tate Etc.*, vol. 10, Summer 2007.
58. George A. Akerlof, "The Missing Motivation in Macroeconomics," presidential address, meeting of the American Economic Association, Chicago, January 6, 2007.
59. Sarah F. Brosnan and Frans B. M. de Waal, "Monkeys Reject Unequal Pay," *Nature*, vol. 425, September 18, 2003, pp. 297–99.
60. Cited in George Ainslie, "Specious Reward: A Behavioral Theory of Impulsiveness and Impulse Control," *Psychological Bulletin*, vol. 82, no. 4, July 1975, p. 476.
61. Ernst Fehr, Urs Fischbacher, and Michael Kosfeld, "Neuroeconomic Foundations of Trust and Social Preferences: Initial Evidence," *American Economic Review*, vol. 95, no. 2, May 2005, pp. 346–51.
62. Joseph Heinrich, Robert Boyd, Samuel Bowles, et al., "'Economic Man' in Cross-Cultural Perspective: Behavioral Experiments in 15 Small-Scale Societies," *Behavioral and Brain Sciences*, vol. 28, 2005, pp. 795–855.
63. Jeffrey Goldberg, Lívia Markóczy, and Lawrence Zahn, "Symmetry and the Illusion of Control as Bases for Cooperative Behavior," *Rationality and Society*, vol. 17, no. 2, 2005, pp. 243–70.
64. Daria Knoch, Alvaro Pascual-Leone, Kaspar Meyer, Valerie Treyer, and Ernst Fehr, "Diminishing Reciprocal Fairness by Disrupting the Right Prefrontal Cortex," *Science*, vol. 314, November 3, 2006, pp. 829–32.
65. Fehr et al., "Neuroeconomic Foundations."
66. Terence C. Burnham, "High-Testosterone Men Reject Low Ultimatum Game Offers," *Proceedings of the Royal Society B*, accepted June 8, 2007; cited in reference to the work of Steven Quartz at the California Institute of Technology; Henriëtte Prast, "Emotionomics. Neuroeconomics," *Wilmott*, vol. 2005, no. 1, 2005, pp. 42–43. See also Colin Camerer's "What Is Neuroeconomics?" http://www.hss.caltech.edu/~camerer/web_material/n.html.
67. James K. Rilling, David A. Gutman, Thorsten R. Zeh, Giuseppe Pagnoni, Gregory S. Berns, and Clinton D. Kilts, "A Neural Basis for Social Cooperation," *Neuron*, vol. 35, July 18, 2002, pp. 395–405.
68. Manfred Ertel and Padma Rao, "Women Are Better with Money," *Der Spiegel*, December 7, 2006.
69. Hadley Cantril and Albert Hastorf, "They Saw a Game: A Case Study," *Journal of Abnormal and Social Psychology*, vol. 49, 1954, pp. 129–34.
70. Linda Babcock and George Loewenstein, "Explaining Bargaining Impasse:

The Role of Self-Serving Biases," *Journal of Economic Perspectives*, vol. 11, no. 1, 1997, pp. 109–126.

71. Robert O. Deaner, Amit V. Khera, and Michael L. Platt, "Monkeys Pay Per View: Adaptive Valuation of Social Images by Rhesus Macaques," *Current Biology*, vol. 15, March 29, 2005, pp. 543–48.
72. Marianne Bertrand, Dean Karlin, Sendhil Mullainathan, Eldar Shafir, and Jonathan Zinman, "What's Psychology Worth? A Field Experiment in the Consumer Credit Market," National Bureau of Economic Research Working Paper no. 11892, December 2005.
73. Bram Van den Bergh and Siegfried Dewitte, "Digit Ratio (2D:4D) Moderates the Impact of Sexual Cues on Men's Decisions in Ultimatum Games," *Proceedings of the Royal Society B*, accepted March 16, 2006.
74. Thomas N. Robinson, Dina L. G. Borzekowski, Donna M. Matheson, and Helena C. Kraemer, "Effects of Fast-Food Branding on Young Children's Taste Preferences," *Archives of Pediatrics and Adolescent Medicine*, vol. 161, no. 18, August 2007, pp. 792–97.
75. Samuel M. McClure, Jian Li, Damon Tomlin, Kim S. Cypert, Latané M. Montague, and P. Read Montague, "Neural Correlates of Behavioral Preference for Culturally Familiar Drinks," *Neuron*, vol. 44, October 14, 2004, pp. 379–87.
76. Robin Goldstein, Johan Almenberg, Anna Dreber, John W. Emerson, Alexis Herschkowitsch, and Jacob Katz, "Do Expensive Wines Taste Better? Evidence from a Large Sample of Blind Tastings," American Association of Wine Economists Working Paper no. 16, April 2008.
77. *The Kingdom*, a piece by British artist Damian Hirst, sold at Sotheby's Bond Street, September 15, 2008. This was the hammer price; the actual cost to the buyer would be higher.
78. Agnès Gault, Yves Meinard, and Frank Courchamp, "Less Is More: Rarity Trumps Quality in Luxury Markets," *Nature Precedings*, March 18, 2008.
79. Daniel T. Gilbert and Michael Gill, "The Monetary Realist," *Psychological Science*, vol. 11, no. 5, September 2000, pp. 394–98.
80. Leaf van Boven and Thomas Gilovich, "To Do or to Have? That Is the Question," *Journal of Personality and Social Psychology*, vol. 85, no. 6, 2003, pp. 1193–1202.
81. Kathleen D. Vohs, Nicole L. Mead, and Miranda R. Goode, "The Psychological Consequences of Money," *Science*, vol. 314, November 17, 2006, pp. 1154–56.

CHAPTER III: TINTED GLASSES

1. *Substantia nigra, striatum, corpi mamillari, zona incerta.*
2. Quoted in "Neuromorphic Microchips," *Scientific American*, April 25, 2005.
3. See J. H. Feldman and D. H. Ballard, "Connectionist Models and Their Properties," *Cognitive Science*, vol. 6, 1982, pp. 205–254.
4. Simon P. Liversedge, Keith Rayner, Sarah J. White, John M. Findlay, and Eu-

gene McSorley, "Binocular Coordination of the Eyes During Reading," *Current Biology*, vol. 16, no. 17, 2006, pp. 1726–29.

5. From a letter to *New Scientist*, no. 2188, May 29, 1999, referring to Dr. Rawlinson's Ph.D. thesis at Nottingham University, *The Significance of Letter Position in Word Recognition*, 1976.
6. Ronald C. Kessler, Emil F. Coccaro, Maurizio Fava, Savina Jaeger, Robert Jin, and Ellen Walters, "The Prevalence and Correlates of DSM-IV Intermittent Explosive Disorder in the National Comorbidity Survey Replication," *Archives of General Psychiatry*, vol. 63, 2006, pp. 669–78.
7. See Judgments of the Supreme Court of Canada, *R. v. Parks, [1992] 2 S.C.R. 871.*
8. Rachel Nowak, "Sleepwalking Woman Had Sex with Strangers," *New Scientist*, October 15, 2004.
9. Paul MacLean, "A Mind of Three Minds: Educating the Triune Brain," *Yearbook of the National Society for the Study of Education*, vol. 77, no. 2, 1978, pp. 308–27.
10. This is evidenced by a remarkable finding from the Max Planck Institute for Human Cognitive and Brain Sciences in Leipzig: that apparently free-will decisions (to tap one button or another at a time of one's choosing) are prefigured in brain activity up to seven seconds before the subject is conscious of making them. See Chun Siong Soon, Marcel Brass, Hans-Jochen Heinze, and John-Dylan Haynes, "Unconscious Determinants of Free Decisions in the Human Brain," *Nature Neuroscience*, vol. 11, no. 5, May 2008, pp. 543–45.
11. Donald T. Stuss, Gordon G. Gallup Jr., and Michael P. Alexander, "The Frontal Lobes Are Necessary for 'Theory of Mind,'" *Brain: A Journal of Neurology*, vol. 124, no. 2, February 2001, pp. 279–86; also, in the same issue, C. Fine, J. Lumsden, and R. J. R. Blair, "Dissociation Between 'Theory of Mind' and Executive Functions in a Patient with Early Left Amygdala Damage," pp. 287–98.
12. According to Ken Nakayama at Harvard. See his home page at www.visionlab.harvard.edu/members/nakayama.
13. Kristin Koch, Judith McLean, Ronen Segev, Michael A. Freed, Michael J. Berry II, Vijay Balasubramanian, and Peter Sterling, "How *Much* the Eye Tells the Brain," *Current Biology*, vol. 16, July 25, 2006, pp. 1428–34.
14. Melvyn A. Goodale and A. David Milner, "Separate Visual Pathways for Perception and Action," *Trends in Neurosciences*, vol. 15, no. 1, January 1992, pp. 20–25.
15. David A. Leopold, Igor V. Bondar, and Martin A. Giese, "Norm-Based Face Encoding by Single Neurons in the Monkey Inferotemporal Cortex," *Nature*, vol. 442, August 3, 2006, pp. 572–75.
16. A. D Milner, D. I Perrett, R. S. Johnston, et al., "Perception and Action in Visual Form Agnosia," *Brain*, vol. 114, 1991, pp. 405–28.
17. M. A. Persinger, "Out-of-Body-Like Experiences Are More Probable in People with Elevated Complex Partial Epileptic-Like Signs During Periods of Enhanced Geomagnetic Activity: A Nonlinear Effect," *Perceptual and Motor Skills*, vol. 80, 1995, pp. 563–69.

18. See Bigna Lenggenhager, Tej Tadi, Thomas Metzinger, and Olaf Blanke, "Video Ergo Sum: Manipulating Bodily Self-Consciousness," *Science*, vol. 317, August 24, 2007, pp. 1096–99; and in the same issue, H. Henrik Ehrsson, "The Experimental Induction of Out-of-Body Experiences," p. 1048.
19. See Robert J. Teunisse, Johan R. Cruysberg, Willibrord H. Hoefnagels, André L. Verbeek, and Frans G. Zitman, "Visual Hallucinations in Psychologically Normal People: Charles Bonnet's Syndrome," *Lancet*, vol. 347, March 1996, pp. 794–97.
20. Joshua New, Leda Cosmides, and John Tooby, "Category-Specific Attention for Animals Reflects Ancestral Priorities, Not Expertise," *Proceedings of the National Academy of Sciences*, vol. 104, no. 42, October 16, 2007, pp. 16598–603.
21. G. Johansson, "Visual Perception of Biological Motion and a Model for Its Analysis," *Perceptual Psychophysics*, vol. 14, 1978, pp. 201–211.
22. See the work of George Mather of the University of Sussex, including G. Mather and L. Murdoch, "Gender Discrimination in Biological Motion Displays Based on Dynamic Cues," *Proceedings of the Royal Society of London B*, vol. 258, 1994, pp. 273–79.
23. P. Rochat, R. Morgan, and M. Carpenter, "Young Infants' Sensitivity to Movement Information Specifying Social Causality," *Cognitive Development*, vol. 12, 1997, pp. 537–61.
24. At Bolam lake; see "Yeti Experts Head North to Hunt the Bolam Beast," *Northern Echo*, January 15, 2003.
25. Hans van Kampen, "The Case of the Lost Panda," *Skeptical Inquirer*, vol. 4, no. 1, Fall 1979, pp. 48–50.
26. Sylvia Wright, "The Death of Lady Mondegreen," *Harper's*, November 1954, pp. 48–51.
27. See Bette Bao Lord, *In the Year of the Boar and Jackie Robinson*, HarperTrophy, New York, 1986, p. 87.
28. All right, we'll tell you: Mick Jagger will never "be your beast of burden," while the Shadow knows what evil lurks "in the hearts of men." You can find a rich vein of misheard lyrics at http://www.kissthisguy.com.
29. For this and other interesting speech effects, see J. S. Pardo and R. E. Remez, "The Perception of Speech," in *Handbook of Psycholinguistics*, 2nd ed., ed. M. Traxler and M. A. Gernsbacher, Academic Press, New York, 2006, pp. 201–48.
30. We wouldn't, of course, dream of mentioning "Stopping by Woods on a Snowy Evening" to the tune of "Hernando's Hideaway."
31. For a fascinating discussion of neural time, see David M. Eagleman, Peter U. Tse, Dean Buonomano, Peter Janssen, Anna Christina Nobre, and Alex O. Holcombe, "Time and the Brain: How Subjective Time Relates to Neural Time," *Journal of Neuroscience*, vol. 25, no. 45, November 9, 2005, pp. 10369–71.
32. Daniel B. Wright and Gary Wareham, "Mixing Sound and Vision: The Interaction of Auditory and Visual Information for Earwitnesses of a Crime Scene," *Legal and Criminological Psychology*, vol. 10, 2005, pp. 103–110.

33. Aurélie Bidet-Caulet, Julien Voisin, Olivier Bertrand, and Pierre Fonlupt, "Listening to a Walking Human Activates the Temporal Biological Motion Area," *NeuroImage*, vol. 28, 2005, pp. 132–39.
34. Quoted in Robert Matthews, "Ultra Low Sound Waves Blamed for Visions, Feelings of Terror," *Sunday Telegraph* (London), June 29, 1998; the full description of the event is in Vic Tandy and Tony R. Lawrence, "The Ghost in the Machine," *Journal of the Society for Psychical Research*, vol. 62, no. 851, 1998.
35. Andrea Berger, Gabriel Tzur, and Michael I. Posner, "Infant Brains Detect Arithmetic Errors," *Proceedings of the National Academy of Sciences of the United States*, vol. 103, no. 33, August 15, 2006, pp. 12649–53.
36. Chou P. Hung, Gabriel Kreiman, Tomaso Poggio, and James J. DiCarlo, "Fast Readout of Object Identity from Macaque Inferior Temporal Cortex," *Science*, vol. 310, November 4, 2005, pp. 863–66.
37. Ron Luciano and David Fisher, *The Umpire Strikes Back*, Bantam Dell, New York, 1982, p. 129.
38. Daniel J. Simons and Christopher F. Chabris, "Gorillas in Our Midst: Sustained Inattentional Blindness for Dynamic Events," *Perception*, vol. 28, 1999, pp. 1059–74.
39. A phrase used repeatedly, to apparently great effect, by defense attorney Johnnie Cochran in his closing statement during the O. J. Simpson murder trial, September 27–28, 1995.
40. The cockpit recording was leaked to the *Sun* newspaper and its validity has not been challenged. The material facts are also available in the British Army's "Board of Inquiry Report into the Death of the Late 25035018 Lance Corporal of Horse Matthew Richard Hull" of May 22, 2004.
41. This and other related discoveries in the neurochemistry of memory come from Eric Kandel's lab at Columbia University.
42. Philip Ziegler, "The World at War," *Spectator*, October 10, 2007.
43. Michelle D. Leichtman and Stephen J. Ceci, "The Effects of Stereotypes and Suggestions on Preschoolers' Reports," *Developmental Psychology*, vol. 31, no. 4, July 1995, pp. 568–78.
44. See *State of New Jersey v. Margaret Kelly Michaels*, Superior Court of New Jersey, Appellate Division, 264 N.J. Super. 579, 625 A.2d 489 (1993).
45. See Gary Cartwright, "The Innocent and the Damned," *Texas Monthly Magazine*, April 1994.
46. See, among other works, Elizabeth F. Lotus, "Make-believe Memories," *American Psychologist*, November 2003, pp. 867–73.
47. Ruthanna Gordon, Nancy Franklin, and Jennifer Beck, "Wishful Thinking and Source Monitoring," *Memory and Cognition*, vol. 33, no. 3, 2005, pp. 418–29.
48. You can read about the many different mechanisms of recall and reconsolidation in Natalie C. Tronson and Jane R. Taylor, "Molecular Mechanisms of Memory Reconsolidation," *Nature Reviews Neuroscience*, vol. 8, April 2007, pp. 262–75.
49. R. Quian Quiroga, L. Reddy, G. Kreiman, C. Koch, and I. Fried, "Invariant

Visual Representation by Single Neurons in the Human Brain," *Nature*, vol. 435, June 23, 2005, pp. 1102–7.

50. Kenneth V. Lanning, *Investigator's Guide to Allegations of "Ritual" Child Abuse*, Behavioral Science Unit, National Center for the Analysis of Violent Crime, Federal Bureau of Investigation, Quantico, VA, 1992.
51. For a more nuanced discussion of how memory is modified by maturity, see Noa Ofen, Yun-Ching Kao, Peter Sokol-Hessner, Heesoo Kim, Susan Whitfield-Gabrieli, and John D. E. Gabrieli, "Development of the Declarative Memory System in the Human Brain," *Nature Neuroscience*, vol. 10, no. 9, September 2007, pp. 1198–205.
52. George Lyttelton and Rupert Hart-Davis, *Lyttelton Hart-Davis Letters*, vol. 2, John Murray, London, 1979, letter of October 27, 1955.
53. "Transcript of a recording of Conversation between President Richard Nixon, H. R. Haldeman, and John Dean, on September 15, 1972, at 5:27 to 6:17 P.M. (first installment)," p. 4. available at http://nixon.archives.gov/forresearchers/find/tapes/watergate/trial/transcripts.
54. Ibid., p. 7.
55. Transcript of a telephone conversation between President Nixon and H. R. Haldeman on April 25, 1973, from 7:46 to 7:53 P.M., p. 3.
56. Transcript of a Dictabelt recording of a telephone conversation between E. Howard Hunt and Charles Colson, November 1972.
57. Quoted by Marvin Kalb in *TV Guide*, March 31, 1984, according to Fred Shapiro and Joseph Epstein, *Yale Book of Quotations*, Yale University Press New Haven, 2006, p. 504. Kalb says the phrase was immortalized on the State Department press room door. McCloskey (later Ambassador) was a master of diplomatic obfuscation—once, having been asked to clarify a point, replying, "For clarity, that ambiguity ought to stay." This Robert McCloskey, by the way, is not the same one who wrote *Make Way for Ducklings*.
58. Matthew D. Lieberman, Naomi I. Eisenberger, Molly J. Crockett, Sabrina M. Tom, Jennifer H. Pfeifer, and Baldwin M. Way, "Putting Feelings into Words: Affect Labeling Disrupts Amygdala Activity in Response to Affective Stimuli," *Psychological Science*, vol. 18, no. 5, 2007, pp. 421–28.
59. Justin Kruger, Nicholas Epley, Jason Parker, and Zhi-Wen Ng, "Egocentrism over E-mail: Can We Communicate as Well as We Think?" *Journal of Personality and Social Psychology*, vol. 89, no. 6, 2005, pp. 925–36.
60. Quoted in Daniel Enemark, "It's All About Me: Why E-mails Are So Easily Misunderstood," *Christian Science Monitor*, May 15, 2006.
61. Michael Morris, Janice Nadler, Terri Kurtzberg, and Leigh Thompson, "Schmooze or Lose: Social Friction and Lubrication in E-mail Negotiations," *Group Dynamics: Theory, Research, and Practice*, vol. 6, no. 1, March 2002, pp. 89–100.
62. Thomas Hardy, "She, to Him, III," 1866; *Wessex Poems and Other Verses*, Harper, New York, 1898.
63. Philip Davis in collaboration with Neil Roberts, Victorina Gonzalez-Diaz, and

Guillaume Thierry, "The Shakespeared Brain," *The Reader*, no. 23, Autumn 2006.

64. See Stefan Koelsch, Elizabeth Kasper, Daniela Sammler, Katrin Schulze, Thomas Gunter, and A. D. Friederici, "Music, Language and Meaning: Brain Signatures of Semantic Processing," *Nature Neuroscience*, vol. 7, no. 3, March 2004, pp. 302–7.
65. Thomas Hobbes, *Leviathan*, Crooke, London, 1651, ch. 6.
66. Norbert Wiener, *Cybernetics*, John Wiley, New York, 1948, p. 155.
67. Sabrina Diano, Susan A. Farr, Stephen C. Benoit, Ewan C. McNay, Ivaldo da Silva, Balazs Horvath, F. Spencer Gaskin, Naoko Nonaka, Laura B. Jaeger, William A. Banks, John E. Morley, Shirly Pinto, Robert S. Sherwin, Lin Xu, Kelvin A. Yamada, Mark W. Sleeman, Matthias H. Tschöp, and Tamas L. Horvath, "Ghrelin Controls Hippocampal Spine Synapse Density and Memory Performance," *Nature Neuroscience*, vol. 9, February 19, 2006, pp. 381–88; Tamas L. Horvath was quoted in Andy Coghlan, "Starve Your Stomach to Feed Your Brain," *New Scientist*, February 25, 2006.
68. X. H., "Wonders of the Microscope," *The Olio; or, Museum of Entertainment*, Joseph Shackell, London, 1829, pp. 411–12.
69. Daniel Pinchbeck, "Primordial Wisdom Teacher," Erowid Experience Vaults, January 18, 2003, http://erowid.org/experiences/exp.php?ID=20492.
70. E. D. Dzoljic, C. D. Kaplan, and M. R. Dzoljic, "Effects of Ibogaine on Naloxone-Precipitated Withdrawal Syndrome in Chronic Morphine-Dependent Rats," *Archive of International Pharmacodynamics*, vol. 294, 1988, pp. 64–70.
71. Patrick K. Kroupa, "The Addiction Series: Breaking the Cycle: Staying Clean," *Heroin Times*, 2002, http://www.herointimes.com/mayo2/intervent.html.
72. R. R. Griffiths, W. A. Richards, M. W. Johnson, U. D. McCann, and R. Jesse, "Mystical-Type Experiences Occasioned by Psilocybin Mediate the Attribution of Personal Meaning and Spiritual Significance 14 Months Later," *Journal of Psychopharmacology*, vol. 22, August 2008, pp. 621–32.
73. Oliver Sacks, "The Dog Beneath the Skin," *An Anthropologist on Mars*, Vintage Books, New York, 1995, chapter 18.

CHAPTER IV: OFF THE RAILS

1. For an example of resistentialism in action, see Megan S. C. Lim, Margaret E. Hellard, and Campbell K. Aitken, "The Case of the Disappearing Teaspoons: Longitudinal Cohort Study of the Displacement of Teaspoons in an Australian Research Institute," *British Medical Journal*, vol. 331, December 24, 2005, pp. 1498–500.
2. From A. Tversky and D. Kahneman, "Judgment Under Uncertainty: Heuristics and Biases," *Science*, vol. 185, 1974, pp. 1124–31.
3. For this and many other amusing illusions, see Massimo Piatelli-Palmarini, *Inevitable Illusions*, John Wiley, New York, 1994, pp. 116–20.

4. Daniel L. Schacter and Donna Rose Addis, "The Optimistic Brain," *Nature Neuroscience*, vol. 10, no. 11, November 2007, pp. 1345–47.
5. See A. Conan Doyle, "Silver Blaze," *The Memoirs of Sherlock Holmes*, 1892:

   > "Is there any point to which you would wish to draw my attention?"
   > "To the curious incident of the dog in the night-time."
   > "The dog did nothing in the night-time."
   > "That was the curious incident," remarked Sherlock Holmes.

6. This test originates in a paper by D. M. Eddy in 1982, but has been extensively explored by Gerd Gigerenzer, recently in Gerd Gigerenzer and Adrian Edwards, "Simple Tools for Understanding Risks: From Innumeracy to Insight," *British Medical Journal*, vol. 327, September 27, 2003, pp. 741–44.
7. This is called the Wason test, by the way. Many commentators have made much of it.
8. Our description is based on the official Department for Transport Aircraft Accident Report no 4/90 (EW/C1095).
9. You will find many more, horrifying, stories of this kind in James R. Chiles, *Inviting Disaster*, HarperCollins, New York, 2001.
10. Our source is "F-22 Squadron Shot Down by the International Date Line," *Defense Industry Daily*, March 1, 2007.
11. Our source for this event is the *Report to the President by the Presidential Commission on the Space Shuttle* Challenger *Accident*, ch. 5, Government Printing Office, Washington, D.C. June 6, 1986.
12. Ibid., vol. 2, appendix F.
13. See Kirk Johnson, "Denver Airport Saw the Future. It Didn't Work," *New York Times*, August 27, 2005.
14. Source: Dave Miller, television cameraman, personal interview.
15. Lisa Feigenson and Justin Halberda, "Conceptual Knowledge Increases Infants' Memory Capacity," *Proceedings of the National Academy of Sciences*, vol. 105, no. 29, July 22, 2008, pp. 9926–30.
16. A good source for these matters is Joshua B. Tenenbaum, Thomas L. Griffiths and Charles Kemp "Theory-based Bayesian Models of Inductive Learning and Reasoning," *Trends in Cognitive Sciences*, vol. 10, no. 7, July 2006, pp. 309–18.
17. Kim Drake and R. Bull, "Understanding Suggestibility: The Link Between Life Adversity and Interviewee Vulnerability," paper presented at the International Congress of Psychology and Law, Adelaide, Australia, July 3–8, 2007.
18. Peter McLeod, Nick Reed, and Zoltan Dienes, "How Fielders Arrive in Time to Catch the Ball," *Nature*, vol. 426, November 20, 2003, pp. 244–45.
19. David Hill, interviewed in Nick T. Spark, *A History of Murphy's Law*, Periscope Film, Los Angeles, 2006.
20. Peter Gay, *Freud for Historians*, Oxford University Press, New York, 1985, p. 77.

21. An excellent departure point for this topic is Philip E. Ross, "The Expert Mind," *Scientific American*, July 24, 2006.
22. Quoted in *Chess Life*, March 1984, p. 26.
23. Quoted in Seiji Nagata, *Hokusai: Genius of the Japanese Ukiyo-e*, Kodansha International, Tokyo, 1995, p. 82.
24. Susan Hockfield and Paul Lombroso, "Development of the Cerebral Cortex: IX. Cortical Development and Experience: I," *Journal of the American Academy of Child and Adolescent Psychiatry*, vol. 37, no. 9, 1998, pp. 992–93.
25. Joshua T. Trachtenberg, Brian E. Chen, Graham W. Knott, Guoping Feng, Joshua R. Sanes, Egbert Welker, and Karel Svoboda, "Long-Term in Vivo Imaging of Experience-Dependent Synaptic Plasticity in Adult Cortex," *Nature*, vol. 420, December 19, 2002, pp. 788–94.
26. Eleanor A. Maguire, David G. Gadian, Ingrid S. Johnsrude, Catriona D. Good, John Ashburner, Richard S. J. Frackowiak, and Christopher D. Frith, "Navigation-Related Structural Change in the Hippocampi of Taxi Drivers," *Proceedings of the National Academy of Sciences*, vol. 87, no. 8, April 11, 2000, pp. 4398–403.
27. Quoted in Lawrence S. Ritter, *The Glory of Their Times*, Harper, New York, 1992, p. 282.
28. Our principal source for this is Major Tony Kern, "Darker Shades of Blue: A Case Study of Failed Leadership," Neil Krey's CRM Developer's Forum, 1995; http://www.crm-dev.org/resources/paper/darkblue/darkblue.htm.
29. Clifford Winston, Vikram Maheshri, and Fred Mannering, "An Exploration of the Offset Hypothesis Using Disaggregate Data: The Case of Airbags and Antilock Brakes," *Journal of Risk and Uncertainty*, vol. 32, 2006, pp. 83–99.
30. You can read James Reason's excellent book *Human Error*, Cambridge University Press, Cambridge, 1990, or you can get a quick summary in his article "Human Error: Models and Management," *British Medical Journal*, vol. 320, 2000, pp. 768–70.
31. There is a good thesis on this subject by Major Joel B. Witte, at the U.S. Airforce Institute of Technology: "An Investigation Relating Longitudinal Pilot-Induced Oscillation Tendency Rating to Describing Function Predictions for Rate-Limited Actuators" (AFIT/GAE/ENY/04-M16).
32. B. F. Skinner, "'Superstition' in the Pigeon," *Journal of Experimental Psychology*, vol. 38, 1947, pp. 168–72.
33. Ziva Kunda, "The Case for Motivated Reasoning," *Psychological Bulletin*, vol. 108, no. 3, 1990, pp. 480–97.
34. Quoted in Spark, *Murphy's Law*, p. 41.
35. S. A. Myers and N. Gleicher, "A Successful Program to Lower Cesarean-Section Rates," *New England Journal of Medicine*, vol. 319, December 8, 1988, pp. 1511–16.
36. For this and other fascinating examples, see George Loewenstein, "Out of

Control: Visceral Influences on Behavior," *Organizational Behavior and Human Decision Processes*, vol. 65, no. 3, March 1996, pp. 272–92.

37. M. Tucker and R. Ellis, "On the Relationship Between Seen Objects and Components of Potential Actions," *Journal of Experimental Psychology: Human Perception and Performance*, vol. 24, 1998, pp. 830–46.
38. For an account of brain activity preceding errors in monotonous tasks, see T. Eichele, S. Debener, V. Calhoun, K. Specht, A. Engel, K. Hugdahl, D. von Cramon, and M. Ullsperger, "Prediction of Human Errors by Maladaptive Changes in Event-Related Brain Networks," *Proceedings of the National Academy of Sciences*, vol. 105, no. 16, April 2008, pp. 6173–78.
39. For a discussion of what the brain is doing during daydreaming, see Malia F. Mason, Michael I. Norton, John D. Van Horn, Daniel M. Wegner, Scott T. Grafton, and C. Neil Macrae, "Wandering Minds: The Default Network and Stimulus-Independent Thought," *Science*, vol. 315, January 19, 2007, pp. 393–95.
40. The originator of this "Shared Spaces" concept is Hans Monderman of the Netherlands; it is well explained in Ben Hamilton-Baillie and Phil Jones, "Improving Traffic Behaviors and Safety Through Urban Design," *Civil Engineering*, vol. 158, May 2005, pp. 39–47.
41. In Gerd Gigerenzer, "I Think, Therefore I Err," *Social Research*, vol. 72, no. 1, Spring 2005, pp. 1–24.
42. For details, see the National Transportation Safety Board Aircraft Accident Report no. PB9-910406 (NTSB/AAR-90/06). The cockpit voice recording transcript begins on p. 21.
43. L. T. Kohn, J. M. Corrigan, M. S. Donaldson, ed., and the Institute of Medicine (U.S.) Committee on Quality of Health Care in America, *To Err Is Human: Building a Safer Health System*, National Academy Press, Washington, D.C., 2000.
44. See Kate Murphy, "What Pilots Can Teach Hospitals About Patient Safety," *New York Times*, October 31, 2006.
45. These and other experiments are discussed in Dietrich Dörner, *The Logic of Failure*, Basic Books, Reading, MA, 1996.
46. See Michael R. Gordon, "Ex-Soviet Pilot Still Insists KAL 007 Was Spying," *New York Times*, December 9, 1996.
47. See David Hoffman, "I Had a Funny Feeling in My Gut," *Washington Post*, February 10, 1999.

CHAPTER V: ONE OF US

1. See in particular the photographs in Guillaume Duran, *Libertate Roumanie 1989*, Editions Denoel, Paris, 1990.
2. At the "battle" of Karánsebes, 1788.
3. See the Middle East Media Research Institute Special Dispatch No. 593, "Panic in Khartoum," October 22, 2003, quoting the Sudanese journalist Ja'afar Abbas writing in the Saudi daily newspaper *Al-Watan*, September 24, 2003.

4. Muzafer Sherif, O. J. Harvey, B. Jack White, William R. Hood, and Carolyn W. Sherif, *Intergroup Conflict and Cooperation: The Robbers Cave Experiment*, originally published 1954, republished by Wesleyan Press, Middletown, CT, 1988.
5. Gili Peleg, Gadi Katzir, Ofer Peleg, Michal Kamara, Leonid Brodsky, Hagit Hel-Or, Daniel Keren, and Eviatar Nevo, "Hereditary Family Signature of Facial Expression," *Proceedings of the National Academy of Sciences*, vol. 103, no. 43, October 24, 2006, pp. 15921–26.
6. Martin Daly and Margo I. Wilson, "Whom Are Newborn Babies Said to Resemble?" *Ethology and Sociobiology*, vol. 3, no. 2, 1982, pp. 69–78.
7. L. M. DeBruine, B. C. Jones, A. C. Little, and D. I. Perrett, "Social Perception of Facial Resemblance in Humans," *Archives of Sexual Behavior*, vol. 37, no. 1, pp. 64–77.
8. For a survey of this work, see Vittorio Gallese, Christian Keysers, and Giacomo Rizzolatti, "A Unifying View of the Basis of Social Cognition," *Trends in Cognitive Sciences*, vol. 8, no. 9, September 2004, pp. 396–403.
9. As V. S. Ramachandran, another mirror neuron enthusiast puts it: "Anytime you watch someone else doing something (or even starting to do something), the corresponding mirror neuron might fire in your brain, thereby allowing you to 'read' and understand another's intentions, and thus to develop a sophisticated 'theory of other minds.'" (See his article "Mirror Neurons and Imitation Learning as the Driving Force Behind 'the Great Leap Forward' in Human Evolution" at www.edge.org.)
10. V. K. Ranganathan, V. Siemionow, J. Z. Liu, V. Sahgal, and G. H. Yue, "From Mental Power to Muscle Power—Gaining Strength by Using the Mind," *Neuropsychologia*, vol. 42, no. 7, 2004, pp. 944–56.
11. Y. W. Cheng, O. J. L. Tzeng, J. Decety, T. Imada, and J. C. Hsieh, "Gender Differences in the Human Mirror System: A Magnetoencephalography Study," *Neuroreport*, vol. 17, no. 11, July 31, 2006, pp. 1115–19.
12. For a general summary, see Vilayanur S. Ramachandran and Lindsay M. Oberman, "Broken Mirrors: A Theory of Autism," *Scientific American*, October 16, 2006, pp. 62–69.
13. Dale J. Langford, Sara E. Crager, Zarrar Shehzad, Shad B. Smith, Susana G. Sotocinal, Jeremy S. Levenstadt, Mona Lisa Chanda, Daniel J. Levitin, and Jeffrey S. Mogil, "Social Modulation of Pain as Evidence for Empathy in Mice," *Science*, vol. 312, no. 5782, June 30, 2006, pp. 1967–70.
14. Michael Tomasello, Josep Call, and Brian Hare, "Chimpanzees Understand Psychological States—the Question Is Which Ones and to What Extent," *Trends in Cognitive Sciences*, vol. 7, no. 4, April 2003, pp. 153–56.
15. Derek E. Lyons, Laurie R. Santos, and Frank C. Keil, "Reflections of Other Minds: How Primate Social Cognition Can Inform the Function of Mirror Neurons," *Current Opinion in Neurobiology*, vol. 16, no. 2, April 2006, pp. 230–34.

16. Alicia P. Melis, Brian Hare, and Michael Tomasello, "Chimpanzees Recruit the Best Collaborators," *Science*, vol. 311, March 3, 2007, pp. 1297–1300.
17. Felix Warneken and Michael Tomasello, "Altruistic Helping in Human Infants and Young Chimpanzees," *Science*, vol. 311, no. 5765, March 3, 2006, pp. 1301–3.
18. The original version of this is in H. Wimmer and J. Perner, "Beliefs About Beliefs: Representation and Constraining Function of Wrong Beliefs in Young Children's Understanding of Deception," *Cognition*, vol. 13, pp. 41–68.
19. Betty M. Repacholi and Alison Gopnik, "Early Reasoning About Desires: Evidence from 14- and 18-Month-Olds," *Developmental Psychology*, vol. 33, no. 1, January 1997, pp. 12–21.
20. He said this to his brother Zoltàn, who had unwisely shouted in English while drowning. Quoted in Michael Korda, *Charmed Lives*, Random House, New York, 1979, p. 163.
21. Robert Frost, "The Death of the Hired Man," *North of Boston*, Henry Holt, New York, 1915.
22. H. Kudo and R. I. M. Dunbar, "Neocortex Size and Social Network Size in Primates," *Animal Behavior*, vol. 62, no. 4, October 2001, pp. 711–22.
23. R. A. Hill and R. I. M. Dunbar, "Social Network Size in Humans," *Human Nature*, vol. 14, no. 1, 2003, pp. 53–72.
24. Alex Mesoudi, Andrew Whiten, and Robin Dunbar, "A Bias for Social Information in Human Cultural Transmission," *British Journal of Psychology*, vol. 97, no. 3, pp. 405–23.
25. Mary Elizabeth Burke and Jessica Collison, "Employee Trust and Organizational Loyalty," Society for Human Resource Management, August 2004.
26. Described in Julianna R. Szekely, "What Do We Hate About the Pirese?" *Magyar Hirlap*, July 20, 2006.
27. Rudyard Kipling, "We and They," *Debits and Credits*, Macmillan, London, 1926.
28. Samuel Bowles, "Group Competition, Reproductive Leveling, and the Evolution of Human Altruism," *Science*, vol. 314, December 8, 2006, pp. 1569–72.
29. See Hilliard Kaplan and Michael Gurven, "The Natural History of Human Food Sharing and Cooperation: A Review and a New Multi-Individual Approach to the Negotiation of Norms," in *Moral Sentiments and Material Interests*, ed. Herbert Gintis, MIT Press, Cambridge, MA, 2005.
30. For a demonstration of how to say "*rødgrød med fløde*" see http://www.youtube.com/watch?v=z8VziyktySo.
31. Jonathan Haidt, Paul Rozin, Clark McCauley, and Sumio Imada, "Body, Psyche, and Culture: The Relationship Between Disgust and Morality," *Psychology and Developing Societies*, vol. 9, 1997, pp. 107–31.
32. Molly Parker Tapias, Jack Glaser, Dacher Keltner, Kristen Vasquez, and Thomas Wickens, "Emotion and Prejudice: Specific Emotions Toward Outgroups," *Group Processes and Intergroup Relations*, vol. 10, no. 1, 2007, pp. 27–39.
33. Alexandra J. Golby, John D. E. Gabrieli, Joan Y. Chiao, and Jennifer L. Eberhardt,

"Differential Responses in the Fusiform Region to Same-Race and Other-Race Faces," *Nature Neuroscience*, vol. 4, 2001, pp. 845–50.

34. Elizabeth A. Phelps and Laura A. Thomas, "Race, Behavior, and the Brain: The Role of Neuroimaging in Understanding Complex Social Behaviors," *Political Psychology*, vol. 24, no. 4, pp. 747–58.
35. Eleanor Roosevelt, *This Is My Story*, Harper and Brothers, New York, 1937.
36. Walter L. Ames, *Police and Community in Japan*, University of California Press, Berkeley, 1981, p. 112, claims that police sources estimate that 70 percent of the Yamaguchi-gumi *yakuza* clan (at least in Okayama prefecture, where he conducted most of his research) were Burakumin. See also Eric Johnston, "Osaka Activist's Arrest Lays Bare Yakuza Ties with 'Burakumin,'" *Japan Times*, July 13, 2006.
37. C. M. Steele and J. Aronson, "Stereotype Threat and the Intellectual Test Performance of African-Americans," *Journal of Personality and Social Psychology*, vol. 69, 1995, pp. 797–811.
38. Steven J. Spencer, Claude M. Steele, and Diane M. Quinn, "Stereotype Threat and Women's Math Performance," *Journal of Experimental Social Psychology*, vol. 35, no. 1, January 1999, pp. 4–28.
39. Joshua Aronson, Michael J. Lustina, Catherine Good, Kelli Keough, Claude M. Steele, and Joseph Brown, "When White Men Can't Do Math: Necessary and Sufficient Factors in Stereotype Threat," *Journal of Experimental Social Psychology*, vol. 35, no. 1, January 1999, pp. 29–46.
40. J. Stone, C. I. Lynch, M. Sjomeling, and J. M. Darley, "Stereotype Threat Effects on Black and White Athletic Performance," *Journal of Personality and Social Psychology*, vol. 77, no. 6, 1999, pp. 1213–27.
41. In a study by Steele's student Julio Garcia, then at Tufts, now at the University of Colorado, as reported in Malcolm Gladwell, "The Art of Failure," *New Yorker*, August 21, 2000.
42. To understand how such relationships can develop, see Naomi I. Eisenberger, Carrie Masten, and Eva Telzer, "Experiencing Racial Discrimination: An fMRI Investigation," presented at the Social and Affective Neuroscience Society annual meeting, Boston, June 7, 2008. The point of her research is that it is *less* stressful to believe you are being spurned or excluded because of race than to believe other people reject you as an individual.
43. Craig Haney, Curtis Banks, and Philip Zimbardo, "Interpersonal Dynamics in a Simulated Prison," *International Journal of Criminology and Penology*, vol. 1, 1973, pp. 69–97.
44. M. J. Raleigh, M. T. McGuire, G. L. Brammer, and A. Yuwiler, "Social and Environmental Influences on Blood Serotonin Concentrations in Monkeys," *Archives of General Psychiatry*, vol. 41, no. 4, April 1984.
45. See, inter alia, Alex Pentland, "Social Dynamics: Signals and Behavior," MIT Media Laboratory Technical Note 579, http://vismod.media.mit.edu/tech-reports/TR-579.pdf.

46. His book, *The Authoritarians*, 2006, is available online at http://home.cc.umanitoba.ca/~altemey; he is also the author of *The Authoritarian Specter*, Harvard University Press, Cambridge, MA, 1996.
47. Ibid., p. 25.
48. See, for instance, Paul L. Harris, "What Do Children Learn from Testimony?" in *Cognitive Bases of Science*, ed. P. Carruthers, M. Siegal, and S. Stich, Cambridge University Press, Cambridge, 2002.
49. Melissa A. Koenig, Fabrice Clément, and Paul L. Harris, "Trust in Testimony: Children's Use of True and False Statements," *Psychological Science*, vol. 10, 2004, pp. 694–98.
50. P. Harris and M. Koenig, "Trust in Testimony: How Children Learn About Science and Religion," *Child Development*, vol. 77, May–June 2006, pp. 505–24.
51. A good thing that no one seems able to find in Swift's works.
52. Not to pick unfairly on Islam, but in admiration of its application of new technology to these issues, see "Ask the Imam" at http://islam.tc/ask-imam/index.php.
53. BBC News, "Dawkins: I'm a Cultural Christian," December 10, 2007.
54. Quoted in "A Sceptic's Testimony," *New Scientist*, February 4, 2006, p. 54.
55. John Berryman, "World-Telegram," in *Homage to Mistress Bradstreet and Other Poems*, Noonday Press, New York, 1973.
56. "Christine Boutin rattrapée par ses propos controversés sur le 11-Septembre," *Le Monde*, July 5, 2007.
57. Edward H. Hagen and Gregory A. Bryant, "Music and Dance as a Coalition Signaling System," *Human Nature*, vol. 14, no. 1, 2003, pp. 21–51.
58. See the work of Mahadev Apte at Duke University, Robert Provine at the University of Maryland, and John Moreall at the University of South Florida.
59. See Konrad Lorenz, "Part and Parcel in Animal and Human Societies," in *Studies in Animal and Human Behavior*, vol. 2. Harvard University Press, Cambridge, MA, 1971, pp. 115–95.
60. Quoted in Bob Traer, "Working for Reconciliation:—The Parents Circle/Families Forum," *Ecumenical Accompaniment Programme in Palestine and Israel Forum*, March 18, 2005; available at www.eappi.org/en/news-events/ea-reports/browse/13.

CHAPTER VI: FRESH OFF THE PLEISTOCENE BUS

1. All this extrapolation is the focus of some controversial and fast-changing science. We are prepared to see significant changes in the numbers, but the essential format remains: very few humans, not all that long ago.
2. According to Hans Eiberg, Jesper Troelsen, Mette Nielsen, Annemette Mikkelsen, Jonas Mengel-From, Klaus W. Kjaer, and Lars Hansen, "Blue Eye Color in Humans May Be Caused by a Perfectly Associated Founder Mutation in a

Regulatory Element Located Within the *HERC2* Gene Inhibiting *OCA2* Expression," *Human Genetics*, vol. 123, no. 2, March 2008, pp. 177–87.

3. P. Ekman and W. V. Friesen, "Constants Across Cultures in the Face and Emotion," *Journal of Personality and Social Psychology*, vol. 17, 1971, pp. 124–29.
4. For a discussion of fears as adaptations, see R. M. Nesse, "Evolutionary Explanations of Emotions," *Human Nature*, vol. 1, no. 3, 1990, pp. 261–89, as well as I. M. Marks, *Fears, Phobias, and Rituals: Panic, Anxiety and Their Disorders*, Oxford University Press, New York, 1987.
5. See H. Clark Barrett and Tanya Behne, "Children's Understanding of Death as the Cessation of Agency: A Test Using Sleep Versus Death," *Cognition*, vol. 96, no. 2, June 2005, pp. 93–108; and H. C. Barrett, "Adaptations to Predators and Prey" in *Handbook of Evolutionary Psychology*, ed. D. M. Buss, Wiley, New York, forthcoming.
6. A. Fernald and P. Kuhl, "Acoustic Determinants of Infant Perception for Motherese Speech," *Infant Behavior and Development*, vol. 10, 1987, pp. 279–93; and D. L. Grieser and P. K. Kuhl, "Maternal Speech to Infants in a Tonal Language: Support for Universal Prosodic Features in Motherese," *Developmental Psychology*, vol. 24, 1988, pp. 14–20.
7. See R. Kaplan and S. Kaplan, *The Experience of Nature: A Psychological Perspective*, Cambridge University Press, New York, 1989; and S. Kaplan, "Environmental Preference in a Knowledge-Seeking, Knowledge-Using Organism," in *The Adaptive Mind*, ed. J. H. Barkow, L. Cosmides, and J. Tooby, Oxford University Press, New York, 1992, pp. 535–52.
8. See Jay Appleton, *The Experience of Landscape*, John Wiley, Chichester, 1995.
9. C. S. Henshilwood, "Fully Symbolic Sapiens Behaviour: Innovation in the Middle Stone Age at Blombos Cave, South Africa," in *Rethinking the Human Revolution*, ed. P. Mellars, K. Boyle, O. Bar-Yosef, and C. Stringer, McDonald Institute for Archaeological Research Monograph series, University of Cambridge Press, Cambridge, October 2007.
10. R. Lee and I. DeVore, ed., *Kalahari Hunter-Gatherers: Studies of the !Kung San and Their Neighbors*, Harvard University Press, Cambridge, MA, 1976. The classic article on this issue is Marshall Sahlins, "The Original Affluent Society," in *Man the Hunter*, ed. R. B. Lee and I. DeVore, Aldine Transaction, Piscataway, NJ, 1968, pp. 85–89.
11. This would be an appropriate point to recommend to you Melvin Konner's wonderful book *The Tangled Wing: Biological Constraints on the Human Spirit*, Owl Books, New York, 2002.
12. Monitoring the Future, studies conducted annually by the Institute for Social Research at the University of Michigan, http://www.monitoringthefuture.org.
13. These figures are calculated from census and other data by the Marriage Project at Rutgers University and summarized in its annual "State of Our Unions" report.

14. Richard R. Peterson, "A Re-Evaluation of the Economic Consequences of Divorce," *American Sociological Review*, vol. 61, June 1996, pp. 528–36.
15. David Schramm, "Individual and Social Costs of Divorce in Utah," *Journal of Family and Economic Issues*, vol. 27, 2006, p. 1.
16. Eunice Yu and Jianguo Liu, "Environmental Impacts of Divorce," *Proceedings of the National Academy of Sciences*, vol. 104, no. 51, December 18, 2007, pp. 20629–634.
17. Gustave Flaubert, *Madame Bovary*, pt. 1, ch. 5.
18. Claus Wedekind, Thomas Seebeck, Florence Bettens, and Alexander J. Paepke, "MHC-Dependent Preferences in Humans," *Proceedings of the Royal Society of London B*, vol. 260, no. 1359, 1995, pp. 245–49.
19. This phenomenon does not apply to users of the contraceptive pill: see Seppo Kuukasjä, C. J. Peter Eriksson, Esa Koskela, Tapio Mappes, Kari Nissinen, and Markus J. Rantala, "Attractiveness of Women's Body Odors Over the Menstrual Cycle: The Role of Oral Contraceptives and Receiver Sex," *Behavioral Ecology*, vol. 15, no. 4, 2004, pp. 579–84. For voice, see R. Nathan Pipitone and Gordon G. Gallup Jr., "Women's Voice Attractiveness Varies Across the Menstrual Cycle," *Evolution and Human Behavior*, vol. 29, 2008. Geoffrey Miller of the University of New Mexico has noted that lap dancers receive more tips when they are mid-cycle: Geoffrey Miller, Joshua M. Tybur, and Brent D. Jordan, "Ovulatory Cycle Effects on Tip Earnings by Lap Dancers: Economic Evidence for Human Estrus?" *Evolution and Human Behavior*, vol. 28, 2007, pp. 375–81.
20. Davendra Singh, "Adaptive Significance of Waist-to-Hip Ratio and Female Physical Attractiveness," *Journal of Personality and Social Psychology*, vol. 65, 1993, pp. 293–307.
21. Viren Swami and Martin J. Tovée, "Does Hunger Influence Judgments of Female Physical Attractiveness?" *British Journal of Psychology*, vol. 97, no. 3, August 2006, pp. 353–63.
22. S. W. Gangestad, R. Thornhill, and R. A. Yeo, "Facial Attractiveness, Developmental Stability, and Fluctuating Asymmetry," *Ethology and Sociobiology*, vol. 15, 1994, pp. 73–85.
23. Susan M. Hughes, Marissa A. Harrison, and Gordon G. Gallup Jr., "The Sound of Symmetry: Voice as a Marker of Developmental Instability," *Evolution and Human Behavior*, vol. 23, no. 3, May 2002, pp. 173–80.
24. Dennis M. Bramble and Daniel E. Lieberman, "Endurance Running and the Evolution of *Homo*," *Nature*, vol. 432, November 18, 2004, pp. 345–52.
25. James R. Roney, Katherine N. Hanson, Kristina M. Durante, and Dario Maestripieri, "Reading Men's Faces: Women's Mate Attractiveness Judgments Track Men's Testosterone and Interest in Infants," *Proceedings of the Royal Society B*, vol. 273, no. 1598, September 7, 2006, pp. 2169–75.
26. D. M. Buss and D. P. Schmitt, "Sexual Strategies Theory: An Evolutionary Perspective on Human Mating," *Psychological Review*, vol. 100, 1993, pp. 204–32.

27. For a general discussion of these options, see Sarah Blaffer Hrdy, "The Optimal Number of Fathers: Evolution, Demography, and History in the Shaping of Female Mate Preferences," *Annals of the New York Academy of Sciences*, vol. 97, no. 1, April 2000, pp. 75–96.
28. D. M. Buss, "The Strategies of Human Mating," *American Scientist*, vol. 82, 1994, pp. 238–49.
29. M. W. Wiederman, "Evolved Gender Differences in Mate Preferences: Evidence from Personal Advertisements," *Ethology and Sociobiology*, vol. 14, 1993, pp. 331–52; and D. T. Kenrick, E. K. Sadalla, G. Groth, and M. R. Trost, "Evolution, Traits, and the Stages of Human Courtship: Qualifying the Parental Investment Model," *Journal of Personality*, vol. 58, 1990, pp. 97–116.
30. Jack N. Fenner, "Cross-Cultural Estimation of the Human Generation Interval for Use in Genetics-Based Population Divergence Studies," *American Journal of Physical Anthropology*, vol. 128, 2005, pp. 415–23.
31. M. R. Cunningham, A. R. Roberts, C-H Wu, A. P. Barbee, and P. B. Druen, "'Their Ideas of Beauty Are, on the Whole, the Same as Ours': Consistency and Variability in the Cross-Cultural Perception of Female Attractiveness," *Journal of Personality and Social Psychology*, vol. 68, 1995, pp. 261–79; see also Doug Jones, *Physical Attractiveness and the Theory of Sexual Selection*, University of Michigan Press, Ann Arbor, 1996.
32. J. H. Langlois and L. A. Roggman, "Attractive Faces Are Only Average," *Psychological Science*, vol. 1, 1990, pp. 115–21.
33. Yes, *Calvin* Coolidge; the basis is an old joke describing when the president and his wife were touring a poultry farm. Mrs. Coolidge, having been told how often a particular rooster mated each day, said, "Tell that to the President." Coolidge, duly told, asked, "Same hen each time?" "Well, no." "Tell *that* to Mrs. Coolidge."
34. Alison Jolly, "Hair Signals," *Evolutionary Anthropology*, vol. 14, no. 1, 2005, p. 5.
35. D. M. Buss and L. A. Dedden, "Derogation of Competitors," *Journal of Social and Personal Relationships*, vol. 50, pp. 559–70.
36. B. A. Gladue and J. J. Delaney, "Gender Differences in Perception of Attractiveness of Men and Women in Bars," *Personality and Social Psychology Bulletin*, vol. 16, 1990, pp. 378–91.
37. Or so says Daniel Gilbert, author of *Stumbling on Happiness*; see his interview in *Forbes*, September 21, 2006.
38. *Cailles à la financière*, adapted from Marie-Antoine Carême.
39. Centers for Disease Control, "State-Specific Prevalence of Obesity Among Adults, United States 2005," *Morbidity and Mortality Weekly Report*, vol. 55, no. 26, 2006, pp. 985–88.
40. Judy Putnam, Jane Allshouse, and Linda Scott Kantorm, "U.S. Per Capita Food Supply Trends: More Calories, Refined Carbohydrates, and Fats," *FoodReview*, vol. 25, no. 3, Winter 2002, Economic Research Service, U.S. Department of Agriculture, pp. 2–15.

41. Starbucks nutrition, http://www.starbucks.com/retail/nutrition_info.asp.
42. See, inter alia, J. H. Price, S. M. Desmond, R. A. Krol, F. F. Snyder, and J. K. O'Connell, "Family Practice Physicians' Beliefs, Attitudes, and Practices Regarding Obesity," *American Journal of Preventative Medicine*, vol. 3, no. 6, November–December 1987, pp. 339–45; R. Hoppe and J. Ogden, "Practice Nurses' Beliefs About Obesity and Weight Related Interventions in Primary Care," *International Journal of Obesity and Related Metabolic Disorders*, vol. 21, no. 2, February 1997, pp. 141–46; D. Maroney and S. Golub, "Nurses' Attitudes Toward Obese Persons and Certain Ethnic Groups," *Perceptual and Motor Skills*, vol. 75, no. 2, October 1992, pp. 387–91; H. Oberrieder, R. Walker, D. Monroe, and M. Adeyanju, "Attitudes of Dietetics Students and Registered Dietitians Toward Obesity," *Journal of the American Dietetic Association*, vol. 95, no. 8, August 1995, pp. 914–16; and P. Blumberg and L. P. Mellis, "Medical Students' Attitudes Toward the Obese and Morbidly Obese," *International Journal of Eating Disorders*, vol. 4, no. 2, 1985, pp. 169–75.
43. See Traci Mann, Janet A. Tomiyama, Erika Westling, Ann-Marie Lew, Barbra Samuels, and Jason Chatman, "Medicare's Search for Effective Obesity Treatments: Diets Are Not the Answer," *American Psychologist*, vol. 62, no. 3, April 2007, pp. 220–33.
44. See D. Craig Willcox, Bradley J. Willcox, Qimei He, Nien-chiang Wang, and Makoto Suzuki, "They Really Are That Old: A Validation Study of Centenarian Prevalence in Okinawa," *Journal of Gerontology: Biological Sciences*, vol. 63A, no. 4, 2008, pp. 338–49, and other research from the Okinawa Centenarian Study at http://www.okicent.org/publications.html.
45. Sabrina Diano et al., "Ghrelin Controls Hippocampal Spine Synapse Density and Memory Performance," *Nature Neuroscience*, vol. 9, no. 3, March 2006, pp. 382–88.
46. For a review of the function of leptin, see Jeffrey M. Friedman and Jeffrey L. Halaas, "Leptin and the Regulation of Body Weight in Mammals," *Nature*, vol. 395, October 22, 1998, pp. 763–70.
47. This hypothesis, originally proposed in 1962 by the geneticist James Neel, has been losing ground. Researchers have not found a significant difference in food availability between hunter-gatherers and agricultural people, nor have they identified any specific DNA sequence that could represent the "thrifty gene." A more accepted explanation is a "thrifty phenotype," in which the disruption of traditional life creates local episodes of food deprivation, increasing the frequency of hyperglycemic pregnancies and thus of inherited insulin resistance; see C. Nicholas Hales and David J. P. Barker, "The Thrifty Phenotype Hypothesis: Type 2 Diabetes," *British Medical Bulletin*, vol. 60, no. 1, 2001, pp. 5–20.
48. J. M. Newman, F. DeStefano, S. E. Valway, R. R. German, and B. Muneta, "Diabetes-Associated Mortality in Native Americans," *Diabetes Care*, vol. 16, no. 1, January 1993, pp. 297–99.
49. J. R. Kaplan, S. B. Manuck, and C. Shively, "The Effects of Fat and Cholesterol

on Social Behavior in Monkeys," *Psychosomatic Medicine*, vol. 53, no. 6, November–December 1991, pp. 634–42; as for people, see Anita S. Wells, Nicholas W. Read, Jonathan D. E. Laugharne, and N. S. Ahluwalia, "Alterations in Mood After Changing to a Low-Fat Diet," *British Journal of Nutrition*, vol. 79, 1998, pp. 23–30.

50. Captain G. F. Lyon, *The Private Journal of Captain GF Lyon of the HMS* Hecla, *During the Recent Voyage of Discovery Under Captain Parry*, J. Murray, London 1824, pp. 131–32.

51. Quoted in Roy F. Baumeister, Todd F. Heatherington, and Dianne M. Tice, *Losing Control: How and Why People Fail at Self-Regulation*, Academic Press, San Diego, 1994, p. 184.

52. Richard Restak, *Poe's Heart and the Mountain Climber: Exploring the Effect of Anxiety on Our Brains and Our Culture*, Harmony Books, New York, 2004.

53. Susan E. Swithers and Terry L. Davidson, "A Role for Sweet Taste: Calorie Predictive Relations in Energy Regulation by Rats," *Behavioral Neuroscience*, vol. 122, no. 1, February 2008, pp. 161–73.

54. See Jean Ferrières, "The French Paradox: Lessons for Other Countries," *BMJ Heart*, vol. 90, no. 1, 2004, pp. 107–11.

55. This point of view is expressed well in Michael Pollan, *In Defense of Food*, Penguin Press, New York, 2008. There is also some scientific basis for it: for example, a study by Suzanne Higgs at the University of Birmingham ("Cognitive Influences on Food Intake: the Effects of Manipulating Memory for Recent Eating," *Physiology and Behavior*, vol. 94, no. 5, August 6, 2008, pp. 734–39) showed that remembering what you *have* eaten, rather than thinking about what you *will* eat, reduces appetite. And if you can't have a meal with conversation, chew each mouthful twenty times; that works too.

56. Reported by Tim Weber, "Jobs for the Children of Globalization," BBC News Web site, January 29, 2008.

57. NYU Child Study Center, Medco 2004 Drug Trend Report.

58. The argument is summarized in Sarah Blaffer Hrdy, *Mother Nature: A History of Mothers, Infants, and Natural Selection*, Pantheon Books, New York 1999.

59. This definitely needs to be learned: the foster children of warm foster parents become warm parents irrespective of the empathetic qualities of their birth parents. See the work on empathy by Carolyn Zahn-Waxler of the National Institutes of Mental Health, Nancy Eisenberg at the University of Arizona, and Janet Strayer of Simon Fraser University.

60. E. R. Sowell, P. M. Thompson, C. J. Holmes, et al., "In Vivo Evidence for Post-Adolescent Brain Maturation in Frontal and Striatal Regions," *Nature Neuroscience*, vol. 2, no. 10, 1999, pp. 859–61; more recent work by Frances E. Jensen and David K. Urion at the Harvard Medical School suggests that the rewiring of the frontal lobes is not actually completed until sometime between ages twenty-five and thirty.

61. Ralph S. Solecki, Rose L. Solecki, and Anagnostis P. Agelarakis, *The Proto-Neolithic Cemetery in Shanidar Cave*, Texas A&M Press, College Station, 2004.
62. William von Hippel and Sally M. Dunlop, "Aging, Inhibition, and Social Inappropriateness," *Psychology and Aging*, vol. 20, no. 3, 2005, pp. 519–52.
63. There is a good discussion of this in Hideaki Terashima, "The Relationships Among Plants, Animals, and Man in the African Tropical Rainforest," *African Study Monographs*, suppl. 27, March 2001, pp. 43–60.
64. Or so claims Lewis R. Binford in *In Pursuit of the Past: Decoding the Archaeological Record*, Thames and Hudson, New York, 1983; for figs, see Mordechai E. Kislev, Anat Hartmann, and Ofer Bar-Yosef, "Early Domesticated Fig in the Jordan Valley," *Science*, vol. 312, June 2, 2006, pp. 1372–74.
65. Garrett Hardin, "The Tragedy of the Commons," *Science*, vol. 162, no. 3859, December 13, 1968, pp. 1243–48.
66. For a thoroughly depressing read, choose Richard Ellis, *The Empty Ocean: Plundering the World's Marine Life*, Island Press, Washington, D.C., 2003.
67. Andrew Strathern, *Ongka: Self Account by a New Guinea Big-man*, Gerald Duckworth and Co., London, 1979.

CHAPTER VII: LIVING RIGHT

1. Rev. Charles Kingsley, *Health and Education*, ch. 1, W. Ibister, London, 1874.
2. Hume brings up the problem in his *Treatise of Human Nature*, Noon and Longman, London, 1739–1740, bk. 3, pt. 1, sec. 1.
3. In G. E. Moore, *Principia Ethica,* 1903, ch. 1, sec. 10.
4. When Field Marshal Mackensen captured Belgrade in 1915, he immediately raised a monument to his defeated enemies, saying, "We fought against an army that we have only heard about in fairy tales."
5. Quoted in "The Adventurous Life," *Time*, May 4, 1959.
6. J. Kiley Hamlin, Karen Wynn, and Paul Bloom, "Social Evaluation by Preverbal Infants," *Nature*, vol. 450, November 22, 2007, pp. 557–60.
7. Samuel Bulter, *Erewhon*, ch. 23, "The Book of the Machines," 1910.
8. Marc Bekoff, "Wild Justice and Fair Play: Animal Origins of Social Morality," lecture at American Association for the Advancement of Science, Dialogue on Science, Ethics, and Religion, October 16, 2003.
9. Domestic dogs are, in fact, particularly good at this: better at, or at least more interested in, inferring human intention than wild primates are. See Brian Hare and Michael Tomasello, "Human-like Social Skills in Dogs?" *Trends in Cognitive Sciences*, vol. 9, no. 9, September 2005, pp. 439–44.
10. Anything by de Waal seems to us worth reading. Much of the material reviewed in the next paragraphs was gleaned from his *Good Natured: The Origins of Right and Wrong in Humans and Other Animals*, Harvard University Press, Cambridge, MA, 1996.

11. Referred to in Frans de Waal, "The Animal Roots of Human Morality," *New Scientist*, October 14, 2006.
12. Dogs are actually better than apes at reading human social signals: see Brian Hare, Michelle Brown, Christina Williamson, and Michael Tomasello, "The Domestication of Social Cognition in Dogs," *Science*, vol. 298, November 22, 2002, pp. 1632–36.
13. These and similar issues are explored in Jane Bybee, ed., *Guilt and Shame in Children*, Academic Press, San Diego, 1997.
14. J. Haidt, S. Koller, and M. Dias, "Affect, Culture, and Morality, or Is It Wrong to Eat Your Dog?" *Journal of Personality and Social Psychology*, vol. 65, 1993, pp. 613–28.
15. Described in J. Haidt, "Elevation and the Positive Psychology of Morality," in *Flourishing: Positive Psychology and the Life Well-Lived*, ed. C. L. M. Keyes and J. Haidt, American Psychological Association, Washington, D.C., 2003, pp. 275–89.
16. Jorge Moll, Frank Krueger, Roland Zahn, Matteo Pardini, Ricardo de Oliveira-Souza, and Jordan Grafman, "Human Fronto–Mesolimbic Networks Guide Decisions About Charitable Donation," *Proceedings of the National Academy of Sciences*, vol. 103, no. 42, October 17, 2006, pp. 15623–28.
17. Vladas Griskevicius, Joshua M. Tybur, Jill M. Sundie, Robert B. Cialdini, Geoffrey F. Miller, and Douglas T. Kenrick, "Blatant Benevolence and Conspicuous Consumption: When Romantic Motives Elicit Strategic Costly Signals," *Journal of Personality and Social Psychology*, vol. 93, no. 1, 2007, pp. 85–102.
18. Oriel Sullivan and Scott Coltrane, "Men's Changing Contribution to Housework and Child Care," a discussion paper on changing family roles prepared for the Conference of the Council on Contemporary Families, University of Illinois, Chicago, April 25–26, 2008.
19. Joshua Coleman and Stephanie Coontz, ed., *Unconventional Wisdom*, a survey of research and clinical findings prepared for the conference of the Council on Contemporary Families, University of Chicago, May 2007, p. 13.
20. Michael Koenigs, Liane Young, Ralph Adolphs, Daniel Tranel, Fiery Cushman, Marc Hauser, and Antonio Damasio, "Damage to the Prefrontal Cortex Increases Utilitarian Moral Judgements," *Nature*, vol. 446, no. 7138, April 19, 2007, pp. 908–11.
21. The principal theorist about the influence of environment on chimpanzee and bonobo behavior is Gottfried Hohmann of the Max Planck Institute for Evolutionary Anthropology, Leipzig; see M. M. Robbins and G. Hohmann, "Primate Feeding Ecology: An Integrative Approach," in *Feeding Ecology of Apes and other Primates*, ed. G. Hohmann, M. M. Robbins, C. Boesch, Cambridge University Press, Cambridge, 2006, pp. 1–13.
22. Kissinger is quoted in DuPre Jones, "The Sayings of Secretary Henry," the *New York Times Magazine*, October 28, 1973.
23. G. Ward and D. Keltner, *Power and the Consumption of Resources*, unpublished

manuscript, 1998; discussed in Dacher Keltner, Deborah H. Gruenfeld, and Cameron Anderson, "Power, Approach, and Inhibition," *Psychological Review*, vol. 110, no. 2, 2003, pp. 265–84.

24. Said Aldous Huxley in *The Devils of Loudun*, ch. 10; his point is germane here: "It was precisely because great men tried to seem more than human," he wrote, "that the rest of the world welcomed any reminder that, in part at least, they were still merely animal."

25. The guru in this field currently is Robert I. Sutton of Stanford, who has summarized his findings in *The No Asshole Rule: Building a Civilized Workplace and Surviving One That Isn't*, Warner Business Books, New York, 2007.

26. Laura Sofield and Susan W. Salmond, "Workplace Violence: A Focus on Verbal Abuse and Intent to Leave the Organization," *Orthopaedic Nursing*, vol. 22, no. 4, July–August 2003, pp. 274–83; Erica Frank, Jennifer S. Carrera, Terry Stratton, Janet Bickel, and Lois Margaret Nora, "Experiences of Belittlement and Harassment and Their Correlates Among Medical Students in the United States: Longitudinal Survey," *British Medical Journal*, vol. 333, no. 682, September 30, 2006.

27. A. C. Edmondson, "Learning from Failure in Health Care: Frequent Opportunities, Pervasive Barriers," *Quality and Safety in Health Care*, vol. 13, no. 6, December 2004, pp. 3–9.

28. This quotation is cited in Bob Woodward, "Secret Reports Dispute White House Optimism," *Washington Post*, October 1, 2006.

29. Quoted in the *Report of the Official Account of the Bombings in London on 7th July 2005*, House of Commons document HC 1087, section 39, Stationery Office, London, May 11, 2006.

30. Marc Sageman, *Understanding Terror Networks*, University of Pennsylvania Press, Philadelphia, 2004.

31. See Benedikt Herrmann, Christian Thöni, and Simon Gächter, "Antisocial Punishment Across Societies," *Science*, vol. 319, no. 5868, March 2008, pp. 1362–67. See also Helen Bernhard, Ernst Fehr, and Urs Fischbacher, "Group Affiliation and Altruistic Norm Enforcement," *American Economic Review*, vol. 96, no. 2, May 2006, pp. 217–21; as well as Karla Hoff, Mayuresh Kshetramade, and Ernst Fehr, "Does Social Exclusion of a Group Undermine Its Trust-Enforcing Mechanisms? Evidence from a Third Party Punishment Experiment," submission to the Brown Bread Conference, March 10, 2008.

32. Bishop Gianfranco Girotti, head of the Apostolic Penitentiary, quoted in Nicola Gori, "Le Nuove Forme del Peccato Sociale," *L'Osservatore Romano*, March 9, 2008.

33. Well discussed by Hershey H. Friedman of the City University of New York on the Jewish Law Web site, http://www.jlaw.com/Articles/geneivatdaat.html.

34. Kohlberg originally developed his scale of moral development in his Ph.D. thesis at the University of Chicago: "The Development of Modes of Thinking and Choices in Years 10 to 16," 1958.

35. This is the "druggist" or "Heinz" dilemma, described in Lawrence Kohlberg, *Essays on Moral Development, Vol. I: The Philosophy of Moral Development*, Harper and Row, San Francisco, 1981.
36. See Carol Gilligan, "In a Different Voice: Women's Conceptions of Self and Morality," *Harvard Educational Review*, vol. 47, no. 4, 1977, pp. 481–517.
37. A. Terracciano, A. M. Abdel-Khalek, N. Adam, et al., "National Character Does Not Reflect Mean Personality Trait Levels in 49 Cultures," *Science*, vol. 310, October 7, 2005, pp. 96–100.
38. In David Hackett Fischer, *Albion's Seed: Four British Folkways in America*, Oxford University Press, New York, 1991.
39. Described in Ruth Benedict, *Patterns of Culture*, Houghton Mifflin, New York, 1934.

Index

A Note on the Authors

Michael and Ellen Kaplan are son and mother, and coauthors of the bestselling *Chances Are . . . : Adventures in Probability*. Michael is an award-winning writer and filmmaker who lives near Edinburgh, Scotland. Ellen is a historian and cofounder of the Math Circle, a program for the exploration and enjoyment of mathematics. She is coauthor, with her husband, Robert, of *The Art of the Infinite: The Pleasures of Mathematics* and *Out of the Labyrinth: Setting Mathematics Free.* She lives in central Massachusetts.